Structural instability

an introductory course on

Structural instability

fundamentals of post-buckling behaviour of structures

held at the Department of Civil Engineering,
University of Surrey, 23–27 September 1973

Course notes edited for publication by W. J. Supple

Published by IPC Science and Technology Press Ltd

Published by IPC Science and Technology Press Ltd, IPC House, 32 High Street, Guildford, Surrey GU1 3EW, England.

Typeset by Mid-County Press, Grosvenor House, 18 The Ridgway, Wimbledon, London SW19; printed by Whitstable Litho, Millstrood Road, Whitstable, Kent; and bound by The Wigmore Bindery Ltd, Balena Close, Poole, Dorset.

ISBN 0 902852 28 0

Contents

1 General introduction

A general goal in technological development is that of efficiency – to do more with less. We see about us the products of this trend resulting in new materials and methods, miniaturization in electronics, an ever-expanding usage of computers and an overall decrease in wastage of resources. In structural mechanics this trend is witnessed by the increasing number of lightweight structural components which are now used. The aircraft and aerospace industries are largely responsible for this development since in these areas of design and construction weight is a major parameter which must be minimized to make a design project feasible let alone economic. Civil engineering has also followed this trend in structures both independently and also by exploiting the analytical knowledge gained in the design of aeronautical structures. We have seen over the past two decades or so the construction of many thin-shell structures and the emergence of spaceframe technology and other lightweight structural forms. As one might have anticipated, the new situation brings with it new problems and in lightweight construction two new worries for the designer are the associated problems of instability and dynamic response. We have a swings-and-roundabouts situation in that a more efficient design in terms of materials used and erection procedures followed may require a much more detailed analysis to ensure safety standards. If efficiency is to be a prime objective of the design engineer then it must be guaranteed that the structural analysis employed will be sufficient to predict the structural response under the design loads and all probable practical loads which may come onto the structure in its lifetime. In the present work we shall be interested in the phenomenon of structural instability and shall point out the types of problem to be anticipated in the design of lightweight structures. Instability is often, but not always, synonymous with collapse. In the past civil engineers have studied collapse of structures using plastic methods of analysis and indeed collapse of any structure will by the implication of large deformations introduce plastic yield in portions of the structure. However, instability may be initiated whilst the structure is still in the fully elastic range, plasticity being generated by the ensuing gross deformations. The elastic buckling of an Euler strut is a well-known example of this. A large amount of research and study has been carried out in recent years on the elastic stability of structures and it is with this that we shall concern ourselves here. The subject of plastic instability is considered to be beyond the scope of the present work which is of an introductory nature bearing in mind that stability is often considered to be an advanced subject in courses on structural theory.

Investigations into elastic stability have tended to fall into one of two classes. The first is that of *analysis* in which methods of analysing specific instability problems related to some specific structural form are developed; these studies are invaluable since they aim to provide solutions to practical problems and supply the designer with data useful for design or for comparison with results from prototype or model tests. The second is that of the *study of phenomena* which define the instability process. This latter is of fundamental importance since it explains the general problem and groups together and classifies otherwise seemingly unrelated structural problems. Furthermore, it provides explanations for the large scatter of experimental results and large differences of observed behaviour which are peculiar to experimental studies of structural instability. Again, it may be strongly argued that a knowledge of the types of structural instability phenomenon which exist must be a necessary pre-requisite to any structural instability analysis, that is, it helps to know what one is looking for. The striking example of this is the concept of *post-buckling behaviour* with which we shall be exclusively concerned in the sequel. The term post-buckling is perhaps a little misleading in that it implies only those events which occur after buckling has taken place and gives one the impression that a study of such structural properties is rather like locking the stable door after the horse has bolted. However, we find that the *whole* buckling process is intimately linked with the post-buckling characteristics of a structure even in the stable pre-buckling range. We should therefore consider that we are studying the total load-deformation response characteristics of a structure and in so doing we find it a very useful convenience to define and delineate these different regimes of behaviour and their inter-relationship. Early structural stability studies which ignored (or more correctly, were in ignorance of) the effects of post-buckling characteristics often produced results which were not in keeping with the results obtained from experiments. It was this large discrepancy between theoretical and experimental results particularly with regard to the buckling of thin shell structures which made stability studies a fascinating and fruitful area of research in structural mechanics. The names of many researchers could be mentioned whose efforts have increased our knowledge of structural instability but that of W. T. Koiter must take prominence. It was he who made the first full study of the effects of post-buckling characteristics on the instability process and much subsequent work was based on his findings.

In the following chapters, the ideas of nonlinearity and post-buckling behaviour are introduced and examined and it is shown that all elastic instability problems fall into one of a small number of categories. However, for various reasons, we may find that apparently simple problems may involve quite complex behaviour. Chapters 2 and 3 deal with phenomenological ideas and attempt to classify the types of elastic instability which may be encountered. Chapters 4, 5 and 6 show how these categories of instability behaviour do predict the buckling properties of frame, plate and shell structures.

The analysis of many modern (especially lightweight) structures is very complex and is only soluble by the use of computers and matrix methods. Such numerical approaches have been applied to the study of structural instability. Without a prior knowledge of some of the post-buckling complexities which may arise in structural stability problems, particularly with regard to multiple solutions, the analyst may be led to erroneous conclusions by misinterpretation of the results gained from an analysis by computer. Again, much effort may be saved if the analyst has some idea as to which post-buckling category his particular structural problem belongs. Chapters 7 and 8 take up these ideas and examine the implications of post-buckling theory on the numerical study of structural instability.

2 An introduction to elastic stability

J. M. T. Thompson*

1 Introduction

The last decade has seen vast strides in our understanding of elastic buckling phenomena in their true nonlinear context, and the present author has just completed an advanced monograph on the subject [1] in collaboration with Dr G. W. Hunt. This covers both distinct and compound branching points, and ends with a study of the potential dangers of structural optimization which is generating ever more severe instabilities in the continuing search for greater structural efficiency.

As an introduction to this expanding field, we shall here outline the basic concepts and distinct critical points of elastic stability with some emphasis on well-known continuum problems as a complement to the recent teaching text of Croll and Walker [2].

2 Strut formulation

Because of its great familiarity, the pin-ended Euler strut serves as the best introduction to elastic bifurcation theory, despite the fact that its response is in some ways not typical of the field.

Let us consider then the simply-supported column of Fig.1 with length l and axial compressive load P. The strut is assumed to be axially incompressible with

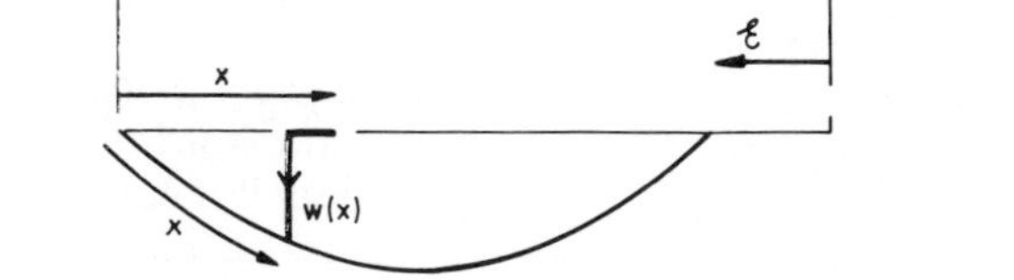

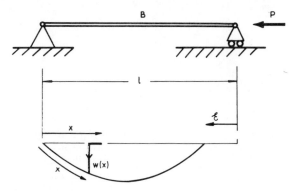

Fig. 1 An inextensible strut suffering large deflections under an axial compressive load.

* Department of Civil Engineering, University College, London.

flexural rigidity $B = EI$, and it will be instructive to start by writing down an exact energy formulation following Thompson and Hunt [1, 3]. Defining carefully the normal displacement $w(x)$ as shown in the figure, the curvature of an element is given by

$$\chi = \frac{d}{dx} \sin^{-1} \dot{w} = \ddot{w}(1 - \dot{w}^2)^{-1/2}, \tag{1}$$

so the strain energy functional is

$$U = \tfrac{1}{2}B \int_0^l \chi^2 \, dx$$

$$= \tfrac{1}{2}B \int_0^l \ddot{w}^2(1 - \dot{w}^2)^{-1} \, dx$$

$$= \tfrac{1}{2}B \int_0^l (\ddot{w}^2 + \ddot{w}^2 \dot{w}^2 + \ddot{w}^2 \dot{w}^4 + \ldots) \, dx. \tag{2}$$

The axial shortening of the column is

$$\mathscr{E} = l - \int_0^l (1 - \dot{w}^2)^{1/2} \, dx$$

$$= \int_0^l (\tfrac{1}{2}\dot{w}^2 + \tfrac{1}{8}\dot{w}^4 + \tfrac{1}{16}\dot{w}^6 + \ldots) \, dx \tag{3}$$

so the total potential energy of the system, structure and load, can be written as

$$V = U - P\mathscr{E}$$

$$= \tfrac{1}{2}B \int_0^l \ddot{w}^2(1 - \dot{w}^2)^{-1} \, dx - P \int_0^l [1 - (1 - \dot{w}^2)^{1/2}] \, dx$$

$$= \tfrac{1}{2}B \underline{\int_0^l (\ddot{w}^2} + \ddot{w}^2\dot{w}^2 + \ddot{w}^2\dot{w}^4 + \ldots)\,\mathrm{d}x - P \underline{\int_0^l (\tfrac{1}{2}\dot{w}^2} + \tfrac{1}{8}\dot{w}^4 + \tfrac{1}{16}\dot{w}^6 \ldots)\,\mathrm{d}x. \tag{4}$$

If, finally, we now use the calculus of variations to find the stationary values of the energy integral we obtain the differential equation of equilibrium

$$\underline{\ddddot{w}[1} + \dot{w}^2 + \dot{w}^4 + \ldots] + 4\dddot{w}\ddot{w}\dot{w}[1 + 2\dot{w}^2 + 3\dot{w}^4 + \ldots] + \ddot{w}^3[1 + 6\dot{w}^2 + 15\dot{w}^4 + \ldots]$$

$$\underline{+ \frac{P}{B}\,\ddot{w}[1} + \tfrac{3}{2}\dot{w}^2 + \tfrac{15}{8}\dot{w}^4 + \ldots] = 0. \tag{5}$$

These exact large-deflection expressions may look rather strange, and for this reason we have underlined the familiar *linear* terms which would appear in a conventional eigenvalue analysis of the engineering texts.

Returning to the complete energy expression it will now be instructive to discretize the problem using a fourier expansion for $w(x)$ and so reduce the total potential energy to a simple algebraic function of the harmonic amplitudes and the load P. Setting then

$$w(x) = \sum_1^\infty u_i \sin \frac{i\pi x}{l}\,, \tag{6}$$

and noting that V will be an even function of the u_i we can write a formal Taylor expansion of the energy

$$V = W = U - P\mathscr{E}$$

$$= \tfrac{1}{2}U_{ij}(0)u_iu_j + \tfrac{1}{24}U_{ijkm}(0)u_iu_ju_ku_m + \ldots - P[\tfrac{1}{2}\mathscr{E}_{ij}(0)u_iu_j$$

$$+ \tfrac{1}{24}\mathscr{E}_{ijkm}(0)u_iu_ju_ku_m + \ldots]. \tag{7}$$

Here we are using subscripts to denote partial differentiation, so that

$$U_{ij}(0) \equiv \left.\frac{\partial^2 U}{\partial u_i \partial u_j}\right|_{u_i=0} \tag{8}$$

and the dummy-suffix summation convention is employed so that summation is always implied over an undefined repeated suffix as

$$\tfrac{1}{2}U_{ij}(0)u_iu_j \equiv \tfrac{1}{2}\sum_i \sum_j U_{ij}(0)u_iu_j\,. \tag{9}$$

Substituting $w(x)$ and its derivatives into the energy functional, the coefficients of the Taylor expansion are readily derived by simple integration of the trigonometric products and can be written as

$$U_{ij}(0) = 0 \text{ for } i \neq j, \qquad U_{ii}(0) = \frac{Bl}{2}\left(\frac{i\pi}{l}\right)^4,$$

$$\mathscr{E}_{ij}(0) = 0 \text{ for } i \neq j, \qquad \mathscr{E}_{ii}(0) = \frac{l}{2}\left(\frac{i\pi}{l}\right)^2,$$

$$U_{1111}(0) = 12B\left(\frac{\pi}{l}\right)^6 \int_0^l \sin^2\frac{\pi x}{l}\cos^2\frac{\pi x}{l}\,\mathrm{d}x$$

$$= {}^3\!/_2 Bl\left(\frac{\pi}{l}\right)^6$$

$$\mathscr{E}_{1111}(0) = 3\left(\frac{\pi}{l}\right)^4 \int_0^l \cos^4\frac{\pi x}{l}\,\mathrm{d}x$$

$$= {}^9\!/_8\left(\frac{\pi}{l}\right)^4 l \qquad \text{, etc.} \tag{10}$$

We now have a *discrete conservative structural system* described by a known algebraic function $V(u_i, P)$ where the u_i are a set of generalized coordinates and P is the loading parameter. Such systems are generated from continuous elastic bodies by *modal* expansions whether the mode-forms be smooth and continuous as in our classical harmonic analysis, or localized and discontinuous [1, 4] as in numerical procedures such as the finite-element method. They are thus of crucial importance to the field of elastic stability and occupy a central place in our recent monograph [1].

3 Some basic principles

Having seen how a discrete system can be generated in familiar surroundings, let us now make some formal developments for a general discrete system described by the algebraic potential energy function $V(Q_i, \Lambda)$ where the Q_i are a set of n generalized coordinates (corresponding to the u_i of the strut) and Λ is some loading parameter (corresponding to P).

For equilibrium of our system at a prescribed value of Λ the energy must be stationary with respect to the generalized coordinates so that for any set of the δQ_i we must have

$$\delta V = V_i \delta Q_i = 0 \tag{11}$$

which of course implies that every first derivative V_i shall be zero. For stability of an equilibrium state at our fixed Λ level the energy must be a complete relative minimum so the second variation of V must not admit negative values and for any set of the δQ_i we must have

$$\delta^2 V = \tfrac{1}{2} V_{ij} \delta Q_i \delta Q_j \geqslant 0. \tag{12}$$

As Λ is subsequently varied, the n equilibrium equations

$$V_i(Q_j, \Lambda) = 0 \tag{13}$$

will generate a series of equilibrium paths in the $(n + 1)$ dimensional load-coordinate space, and it is with the form and stability of these paths that we shall be concerned. At any prescribed load level of $\Lambda = \Lambda^E$ we have seen that the equilibrium and sta-

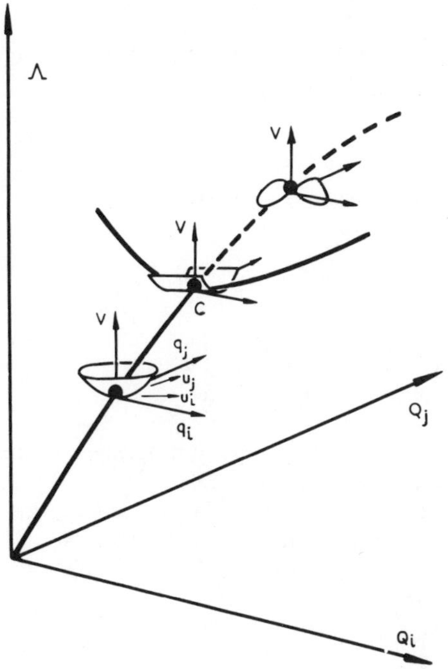

Fig. 2 Equilibrium paths in the n + 1 dimensional load-coordinate space showing loss of stability at a branching point. Energy surfaces are drawn at three distinct load levels.

bility depend on the form of the energy surface $V(Q_i, \Lambda^E)$, and we are led to consider such surfaces at various points on an equilibrium path as shown in Fig.2. We can indeed usefully think of the system as a ball rolling on these energy surfaces.

In this schematic figure we have shown an equilibrium path rising from the unloaded state $\Lambda = Q_i = 0$. This path is supposed to be initially stable, so the first V surface is drawn in the form of a cup. If we now suppose that under increasing Λ this equilibrium path is becoming progressively less stable the curvature of the cup in one (or more) direction will be continuously falling until at some critical point C we shall have *locally* a cylindrical surface along which the ball could be displaced without change of energy. At higher unstable points on the path the V surface will usually still curve upwards in some directions but will curve downwards in others to give a saddle-point, so that under small disturbances our conceptual ball would roll away from the unstable equilibrium state.

Now it is a basic theorem of elastic stability [1, 5] that a rising path cannot just become unstable in this fashion without intersecting a second distinct post-buckling equilibrium path, so we have drawn such a path curving upwards as we shall next observe for the Euler strut.

4 The elastica

Returning to our analysis of the pin-ended strut, we have of course a purely trivial fundamental equilibrium path given by

$$u_i = 0 \text{ for all } P, \tag{14}$$

and the second variation of the total potential energy about a point on this path is

$$\begin{aligned}\delta^2 V &= \tfrac{1}{2} \sum_i \left[U_{ii}(0) - P\mathscr{E}_{ii}(0)\right] u_i^2 \\ &= \tfrac{1}{2} \sum_i \left[\frac{Bl}{2}\left(\frac{i\pi}{l}\right)^4 - P\frac{l}{2}\left(\frac{i\pi}{l}\right)^2\right] u_i^2.\end{aligned} \tag{15}$$

Because of the well-known orthogonality properties of the Fourier integrals we see that $\delta^2 V$ is diagonal and consists simply of a sum of squares. We can thus identify the u_i as a set of *principal* coordinates, and the sin $i\pi x/l$ as the *buckling modes*. The critical load associated with the ith harmonic is obtained by equating to zero the *stability coefficient* multiplying $\tfrac{1}{2}u_i^2$ and is thus simply

$$P^i = \frac{U_{ii}(0)}{\mathscr{E}_{ii}(0)} = B\left(\frac{i\pi}{l}\right)^2. \tag{16}$$

The lowest critical load at which $\delta^2 V$ ceases to be positive definite corresponds to $i = 1$ and is given by the well-known expression

$$P^C = B\left(\frac{\pi}{l}\right)^2 = \frac{\pi^2 EI}{l^2}\ . \tag{17}$$

At this point we might observe that within an eigenvalue analysis the vanishing of the first stability coefficient

$$C_1 = U_{11}(0) - P\mathscr{E}_{11}(0) \tag{18}$$

can be viewed alternatively as the vanishing of V itself, the vanishing of its first derivative V_1 or the vanishing of its second derivative V_{11} corresponding to the three (essentially identical) methods of analysis:

$$V = \tfrac{1}{2}C_1 u_1^2 = 0, \quad \text{Timoshenko's method;}$$

$$V_1 = C_1 u_1 = 0, \quad \text{equilibrium method;}$$

$$V_{11} = C_1 = 0, \quad \text{stability method.} \tag{19}$$

To study the equilibrium solutions in the vicinity of the critical equilibrium state $u_i = 0$, $P = P^C$, we shall anticipate the fact that u_1 will have a higher order of magnitude than u_s where $s \neq 1$ and write the approximate truncated energy function

$$\begin{aligned} V = {} & \underline{\tfrac{1}{2}U_{ii}(0)u_i^2} + \tfrac{1}{24}U_{1111}(0)u_1^4 \\ & - P\left\{\tfrac{1}{2}\mathscr{E}_{ii}(0)u_i^2 + \tfrac{1}{24}\mathscr{E}_{1111}(0)u_1^4\right\}. \end{aligned} \tag{20}$$

The equilibrium equations are then

$$\begin{aligned} \frac{\partial V}{\partial u_1} = {} & \underline{U_{11}(0)u_1} + \tfrac{1}{6}U_{1111}(0)u_1^3 \\ & \underline{- P\{\mathscr{E}_{11}(0)u_1} + \tfrac{1}{6}\mathscr{E}_{1111}(0)u_1^3\} = 0 \\ \frac{\partial V}{\partial u_s} = {} & \underline{U_{ss}(0)u_s - P\mathscr{E}_{ss}(0)u_s} = 0, \end{aligned} \tag{21}$$

where we have again underlined the terms of an engineering eigenvalue analysis, the relevant solution of which would be

$$u_s = 0 \text{ for } s \neq 1$$

$$u_1 = 0 \text{ or } P = P^C \tag{22}$$

with u_1 indeterminate at $P = P^C$. In the higher order post-buckling solution we still have

$$u_s = 0 \text{ for } s \neq 1 \tag{23}$$

but writing

$$P = P^C + p \tag{24}$$

the non-trivial equation for u_1 is now

$$\tfrac{1}{6}\left\{U_{1111}(0) - P\mathscr{E}_{1111}(0)\right\} u_1^2 - p\mathscr{E}_{11}(0) = 0 \tag{25}$$

giving approximately

$$p = \frac{U_{1111}(0) - P^C\mathscr{E}_{1111}(0)}{6\mathscr{E}_{11}(0)} u_1^2$$

$$= \tfrac{1}{2}u_1^2\left[\frac{B}{4}\left(\frac{\pi}{l}\right)^4\right] \tag{26}$$

Here then is a first nonlinear approximation to the upwards curving post-buckling equilibrium path passing through the first critical point.

5 Energy transformations

We return now to our general study of the $V(Q_i, \Lambda)$ system and suppose that a *fundamental* equilibrium path rises monotonically from the unloaded state as shown in Fig.2.

Writing this path in the parametric form

$$Q_i = Q_i^F(\Lambda) \tag{27}$$

we now introduce *incremental* coordinates q_i measured always from this path by means of the equations

$$Q_i = Q_i^F(\Lambda) + q_i. \tag{28}$$

Substituting these into the total potential energy function V gives us this energy in terms of the q_i and we correspondingly denote it by the new symbol W where

$$W(q_i, \Lambda) \equiv V[Q_i^F(\Lambda) + q_i, \Lambda]. \tag{29}$$

This simple transformation is *most important* and should not be dismissed lightly as it can give rise to a great deal of confusion. It was not needed in our discussion of the inextensional strut because in that case the fundamental path was purely trivial and given by $u_i = 0$ for all P.

The first thing to notice about this *sliding* transformation is that it will normally destroy any linearity in Λ. Thus although in many problems V will be a linear function of the loading parameter Λ (as it was in the case of the strut) we must not assume that W will therefore automatically be a linear function of Λ even though in many simple problems this linearity may be preserved.

The second thing to notice is that in an *extensional* problem the corresponding deflection of a simple middle-surface compressive load will be a *linear* function of the in-plane displacement u. This means that V will contain a term of the form $-\Lambda u$. However the fundamental path will now be not purely trivial, and the sliding transformation will generate a W function containing a term of the form $-\Lambda w^2$, where w is a lateral displacement. Thus it is the sliding transformation that brings Λ to act on a *quadratic* function of the normal displacement w as was immediately the case for the inextensional strut. This most important point will be illustrated in the next section for the buckling of a circular plate.

Returning to the general theory, we see that the stability of the fundamental path is now governed by the quadratic form

$$\delta^2 W = \tfrac{1}{2} W_{ij}(0, \Lambda) q_i q_j \tag{30}$$

and it is often convenient to introduce a further coordinate transformation to diagonalize this. The necessary transformation will in the most general case be a function of Λ and can be written as

$$q_i = a_{ij}(\Lambda) u_j. \tag{31}$$

Here the u_i are our new *principal* coordinates, and in terms of these we define the new energy function

$$\begin{aligned} A(u_i, \Lambda) &\equiv W[a_{ij}(\Lambda)u_j, \Lambda] \\ &\equiv V[Q_i^F(\Lambda) + a_{ij}(\Lambda)u_j, \Lambda]. \end{aligned} \tag{32}$$

In some problems $\delta^2 W$ will be linear in Λ and can be written as

$$\delta^2 W = \tfrac{1}{2}(W_{ij}^0 + \Lambda W_{ij}^1) q_i q_j \tag{33}$$

and in this case we can simultaneously diagonalize W_{ij}^0 and W_{ij}^1 by means of a linear transformation which does not depend on Λ and is just a special case of our general diagonalizing transformation.

Stability is now dependent on

$$\delta^2 A = \tfrac{1}{2} A_{ij}(0, \Lambda) u_i u_j, \quad i = j, \tag{34}$$

which is simply a sum of squares and can be written as

$$\delta^2 A = \tfrac{1}{2} C_i(\Lambda) u_i^2 \tag{35}$$

to give us the load-dependent *stability coefficients* C_i. We observe that in the strut problem we were lucky to obtain a diagonal energy function directly because the fourier terms just happened to be the buckling modes.

We use these stability coefficients to make *qualified* statements about the stability of the equilibrium state. Thus if a particular stability coefficient C_r is positive we shall say that the equilibrium state is stable *with respect to* the corresponding principal coordinate u_r. If a particular stability coefficient C_s is negative we shall say that the equilibrium state is unstable *with respect to* the corresponding principal coordinate u_s. If a particular stability coefficient C_t is zero we shall say that the equilibrium state is *critical with respect to* the corresponding principal coordinate u_t.

The number of negative stability coefficients is called the *degree of instability*.

An equilibrium state is called *critical* without qualification if it is critical with respect to one or more of the principal coordinates, and we see that the two determinants $|V_{ij}|$ and $|W_{ij}|$ will vanish in a critical equilibrium state.

If the lowest stability coefficient is positive the equilibrium state is stable with respect to all of the principal coordinates and $\delta^2 V$ is *positive definite*. The equilibrium state is then clearly stable, and in contrast to critically stable equilibrium states whose stability depends on higher variations of V, we might say that the state is *thoroughly stable*.

If the lowest stability coefficient is negative the equilibrium state is unstable with respect to one or more of the principal coordinates and $\delta^2 V$ *admits negative values*. The equilibrium state is then clearly unstable, and in contrast to the critically unstable equilibrium states whose instability depends on higher variations of V, we might say that the state is *thoroughly unstable*.

If the lowest stability coefficient is zero, the equilibrium state is critical with respect to one or more of the principal coordinates and stable with respect to the rest. The quadratic form $\delta^2 V$ is then *positive semi-definite* and supplies no decision about the stability of the equilibrium state. Higher-order terms in the expansion of V must then be examined, but we shall not here pursue this discussion any further.

In discussing the nonlinear phenomena of elastic stability it will often be convenient to think in terms of the diagonalized A function that we have defined, but numerical analysis can advantageously be based on the non-diagonal W energy function which we shall shortly employ in a general branching analysis [1].

6 The circular plate

Before presenting our general W analysis of initial post-buckling we pause for a moment to examine the energy functions of a compressed circular plate to emphasize the distinction between the V and W forms and to observe some necessary properties of the latter.

We consider a thin circular plate of radius R and thickness t deforming symmetrically with respect to its axis of revolution. The movement of a point on the middle surface is decomposed into an inplane displacement $u(r)$ and a normal displacement $w(r)$ as shown in Fig.3, where r is the original distance of the point from

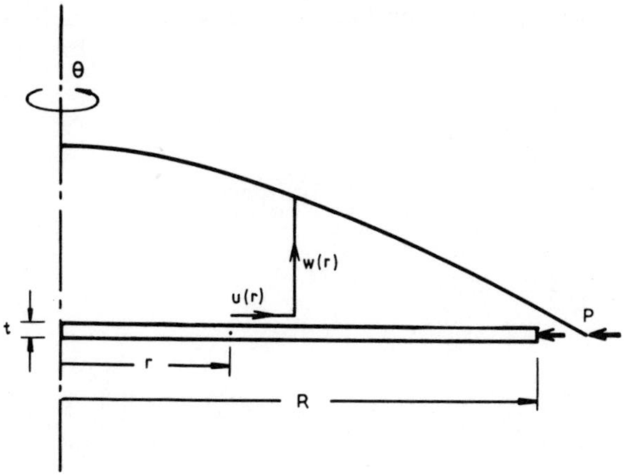

Fig. 3 A circular plate suffering large rotationally-symmetric deflections under a uniform radial compressive loading.

the axis of symmetry. The second independent variable θ representing rotation about this axis will make little appearance in the analysis due to the assumed symmetry. The material of the plate is assumed to be linearly elastic with Young's modulus E and Poisson's ratio ν, and the bending rigidity of the plate is written as

$$D = \frac{Et^3}{12(1-\nu^2)} \; . \tag{36}$$

For moderately large deflections, adequate expressions for the middle-surface curvatures are

$$\chi_r = \frac{d^2w}{dr^2} = \ddot{w}, \quad \chi_\theta = \frac{1}{r}\frac{dw}{dr} = \frac{\dot{w}}{r} \tag{37}$$

where a dot denotes differentiation with respect to r, so the strain energy of bending can be written as

$$J_B = \tfrac{1}{2}D\int(\chi_r^2 + 2\nu\chi_r\chi_\theta + \chi_\theta^2)dA$$

$$= \pi D\int_0^R\left[\ddot{w}^2 + 2\nu\ddot{w}\frac{\dot{w}}{r} + \left(\frac{\dot{w}}{r}\right)^2\right]rdr. \tag{38}$$

Adequate expressions for the middle-surface strains are

$$\epsilon_r = \dot{u} + \tfrac{1}{2}\dot{w}^2, \quad \epsilon_\theta = \frac{u}{r} \tag{39}$$

so the strain energy of stretching can be written as

$$J_E = \frac{Et}{2(1-\nu^2)} \int (\epsilon_r^2 + 2\nu\epsilon_r\epsilon_\theta + \epsilon_\theta^2)\mathrm{d}A$$

$$= \frac{\pi Et}{1-\nu^2} \int_0^R \left[(\dot{u} + \tfrac{1}{2}\dot{w}^2)^2 + 2\nu(\dot{u} + \tfrac{1}{2}\dot{w}^2)\frac{u}{r} + \left(\frac{u}{r}\right)^2 \right] r\mathrm{d}r. \tag{40}$$

In these expressions dA represents an element of area of the middle surface.

The plate is assumed to be symmetrically compressed as indicated by a uniformly distributed force around the circumference. The force system is assumed to be conservative, and such that its potential energy can be written simply as [6]

$$J_L = 2\pi RPu(R) \tag{41}$$

where P is the force per unit length.

The total potential energy of the system, plate and loading device, can finally be written as

$$V = J_B + J_E + J_L$$

$$= \pi D \int_0^R \left[\ddot{w}^2 + 2\nu\ddot{w}\frac{\dot{w}}{r} + \left(\frac{\dot{w}}{r}\right)^2 \right] r\mathrm{d}r + \frac{\pi Et}{1-\nu^2} \int_0^R \left[(\dot{u} + \tfrac{1}{2}\dot{w}^2)^2 \right.$$

$$\left. + 2\nu(\dot{u} + \tfrac{1}{2}\dot{w}^2)\frac{u}{r} + \left(\frac{u}{r}\right)^2 \right] r\mathrm{d}r + 2\pi RPu(R). \tag{42}$$

Now prior to buckling we shall have the simple fundamental response given by

$$w(r) = 0 \quad u(r) = u^F(r) = -\frac{P(1-\nu)}{Et} r \tag{43}$$

and we now define changes in the displacements from this state by the equations

$$w(r) = w(r)$$

$$u(r) = u^F(r) + v(r). \tag{44}$$

This corresponds to the sliding transformation of the discrete coordinate general theory, and we can now write the total potential energy of the system as a function of these incremental displacements,

$$W = \pi D \int_0^R \left[\ddot{w}^2 + 2\nu\ddot{w}\frac{\dot{w}}{r} + \left(\frac{\dot{w}}{r}\right)^2 \right] r\mathrm{d}r$$

$$+ \frac{\pi E t}{1-\nu^2} \int_0^R \left[(\dot{u}^F + \dot{v} + \tfrac{1}{2}\dot{w}^2)^2 + 2\nu(\dot{u}^F + \dot{v} + \tfrac{1}{2}\dot{w}^2)\frac{u^F + v}{r} \right.$$

$$\left. + \left(\frac{u^F + v}{r}\right)^2 \right] r\mathrm{d}r + 2\pi R P[u^F(R) + v(R)]. \tag{45}$$

Ignoring any constant terms, and noting that the linear terms vanish because the fundamental state is one of equilibrium, this can be simplified to

$$W(v, w, P) = \pi D \int_0^R \left[\ddot{w}^2 + 2\nu\ddot{w}\frac{\dot{w}}{r} + \left(\frac{\dot{w}}{r}\right)^2 \right] r\mathrm{d}r$$

$$+ \frac{\pi E t}{1-\nu^2} \int_0^R \left[(\dot{v} + \tfrac{1}{2}\dot{w}^2)^2 + 2\nu(\dot{v} + \tfrac{1}{2}\dot{w}^2)\frac{v}{r} + \left(\frac{v}{r}\right)^2 \right] r\mathrm{d}r$$

$$- \pi P \int_0^R \dot{w}^2 r\mathrm{d}r. \tag{46}$$

We notice that this has exactly the same form as $V[u, w, P]$ with the linear load term replaced by a quadratic load term.

If we now perform a finite-element analysis of this plate [6] with generalized coordinates a_i associated with the incremental in-plane displacement $v(r)$ and coordinates b_j associated with the normal displacement $w(r)$, we see by inspection that we shall obtain our transformed total potential energy function W in the form

$$W(q_i, P) = \tfrac{1}{2}J_{ij}b_ib_j + \tfrac{1}{2}K_{ij}a_ia_j + \tfrac{1}{2}L_{ijk}b_ib_ja_k + \tfrac{1}{24}M_{ijkl}b_ib_jb_kb_l - \tfrac{1}{2}PN_{ij}b_ib_j \qquad (47)$$

where the coordinates q_i embrace both the a_i and the b_j, and the J_{ij} etc, are constants which can be evaluated. We see that W contains quadratic, cubic and quartic terms but no linear terms since the state $a_i = b_j = 0$ is one of equilibrium for all P.

The stability of the fundamental uniformly-compressed state depends on the quadratic form

$$\delta^2W = \tfrac{1}{2}W_{ij}(0, P)q_iq_j = \tfrac{1}{2}J_{ij}b_ib_j + \tfrac{1}{2}K_{ij}a_ia_j - \tfrac{1}{2}PN_{ij}b_ib_j \qquad (48)$$

and with the observed segregation of the a_i and the b_j we see that the critical loads will be given by the vanishing of the sub-determinant

$$|J_{ij} - PN_{ij}| \qquad (49)$$

An initial post-buckling analysis can be written down in terms of the cubic and quartic coefficients [6] but we shall not pursue this example further and return now to our general developments.

7 Perturbation theory

As we have seen with the Euler strut we can expect a post-buckling equilibrium path to emerge from a critical equilibrium state lying on a rising fundamental path, and we present now a form of *perturbation* analysis which has proved invaluable in making local nonlinear studies of such a path [1]. This asymptotic approach seeks to construct a Taylor series for the path about the given critical equilibrium state. It can be based on either the W or A functions, and we shall here employ the former which results in a more general and more elegant treatment.

Returning then to our general transformed energy function $W(q_i, \Lambda)$ relating to a given fundamental equilibrium path, we now seek a post-buckling equilibrium path emerging from the lowest critical equilibrium state in the parametric form

$$q_j = q_j(q_1), \quad \Lambda = \Lambda(q_1), \qquad (50)$$

where the first generalized coordinate q_1 is assumed to be a suitable expansion parameter which partakes of the instability. Substituting these parametric forms into the equilibrium equations $W_i = 0$ gives us the identity

$$W_i[q_j(q_1), \Lambda(q_1)] \equiv 0, \qquad (51)$$

and we can differentiate this repeatedly with respect to q_1 to obtain the ordered set of sequentially linear equilibrium equations

$$W_{ij}q_j^{(1)} + W_i'\Lambda^{(1)} = 0$$

$$(W_{ijk}q_k^{(1)} + W_{ij}'\Lambda^{(1)})q_j^{(1)} + W_{ij}q_j^{(2)}$$

$$+ (W_{ij}'q_j^{(1)} + W_i''\Lambda^{(1)})\Lambda^{(1)} + W_i'\Lambda^{(2)} = 0 \tag{52}$$

etc. Here

$$q_j^{(r)} \equiv \frac{d^r q_j}{dq_1^r}, \quad \Lambda^{(r)} \equiv \frac{d^r\Lambda}{dq_1^r} \tag{53}$$

so that

$$q_1^{(r)} = \delta_{1r} \tag{54}$$

where δ_{ij} is the Kronecker Delta. Subscripts on W denote differentiation with respect to the corresponding generalized coordinates, and a prime denotes differentiation with respect to Λ: the dummy-suffix summation convention is employed with all summations ranging over the n coordinates.

The fundamental path being one of equilibrium, we have $W_i^C = W_i'^C = W_i''^C = \ldots = 0$, where C denotes evaluation at the critical state $q_j = 0$, $\Lambda = \Lambda^C$, so evaluation of our equilibrium equations at this point yields

$$W_{ij}q_j^{(1)}|^C = 0$$

$$W_{ijk}q_j^{(1)}q_k^{(1)} + 2W_{ij}'\Lambda^{(1)}q_j^{(1)} + W_{ij}q_j^{(2)}|^C = 0, \text{ etc.} \tag{55}$$

Remembering that $|W_{ij}^C| = 0$ and $q_1^{(1)} = 1$ we see that the first set of equations can be solved for the rates $q_j^{(1)C}$.

Multiplying the ith equation of the second set of equilibrium equations by $q_i^{(1)C}$ and adding the resulting equations gives us

$$W_{ijk}q_i^{(1)}q_j^{(1)}q_k^{(1)} + 2W_{ij}'q_i^{(1)}q_j^{(1)}\Lambda^{(1)} + W_{ij}q_i^{(1)}q_j^{(2)}|^C = 0, \tag{56}$$

and noting that the last term vanishes by virtue of the first-order equations, we have the contracted equation

$$\Lambda^{(1)C} = -\left.\frac{W_{ijk}q_i^{(1)}q_j^{(1)}q_k^{(1)}}{2W_{ij}'q_i^{(1)}q_j^{(1)}}\right|^C. \tag{57}$$

We see that we have here an expression for the initial slope of the post-buckling equilibrium path on a plot of the load Λ against our first generalized coordinate q_1

and since we have no reason to suppose that the numerator will be zero we observe that we must in general expect a non-zero slope. This is in contrast to our experience with the Euler strut where this slope vanishes due to the physical symmetry of the system.

When the slope is non-zero at a distinct critical point we shall call the branching point an *asymmetric point of bifurcation* [7] and for such a point we have already obtained the first-order nonlinear response. If however the numerator and hence the slope should vanish we can continue the perturbation scheme and the uncontracted second set of equations can be solved for the rates $q_j^{(2)C}$ while the third set of equations can be contracted to yield the second load derivative

$$\Lambda^{(2)C} = -\left.\frac{W_{ijkl}q_i^{(1)}q_j^{(1)}q_k^{(1)}q_l^{(1)} + 3W_{ijk}q_i^{(1)}q_j^{(1)}q_k^{(2)}}{3W'_{ij}q_i^{(1)}q_j^{(1)}}\right|^C \tag{58}$$

When this post-buckling path curvature is positive we shall say that we have a *stable-symmetric* point of bifurcation as with the Euler strut, while if the curvature is negative we shall say that we have an *unstable-symmetric* point of bifurcation.

8 Distinct critical points

The most general way in which a discrete structural system can lose its stability is at a local maximum or *limit point* as shown in Fig.4, but this critical point is well-known in the response of shallow arches and domes and we shall not elaborate on it further.

Of more interest are the three distinct branching points of Figs 5, 6 and 7, and to

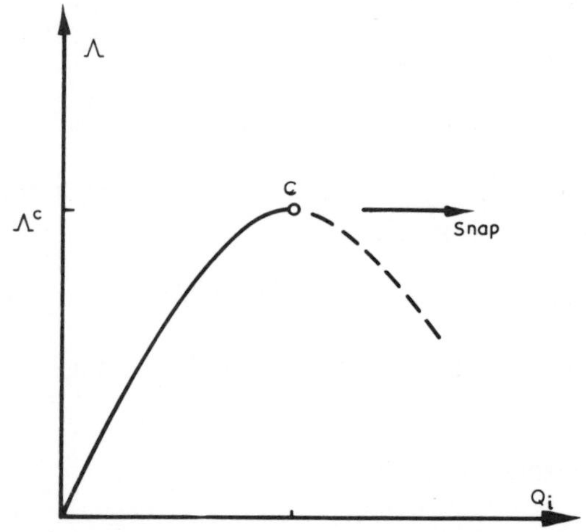

Fig.4 The limit point as associated with the snap-through of shallow arches and domes.

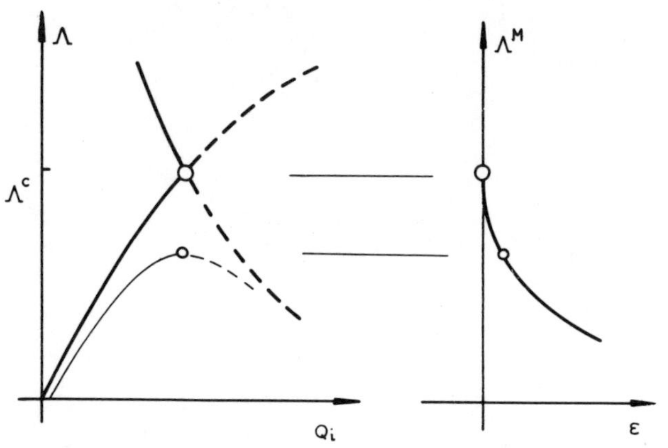

Fig.5 The asymmetric point of bifurcation exhibiting imperfection-sensitivity for 'positive' initial imperfections.

discuss these in a meaningful way we must recognize that physical systems will inevitably contain *initial imperfections.* The influence of these can be studied in a general perturbation approach [1, 7] by considering a total potential energy function $V(Q_i, \Lambda, \epsilon)$ where ϵ is now an imperfection parameter, but we shall here only discuss this influence in a qualitative way.

In the three figures heavy lines represent the equilibrium paths of some perfect system, while light lines represent the paths of equivalent imperfect systems. Continuous lines denote stable equilibrium paths while broken lines denote unstable equilibrium paths.

For the asymmetric point of bifurcation of Fig.5 we see that in the left-hand equilibrium paths the initial imperfections can have a serious effect on the load-carrying capacity of the structure giving us the *imperfection-sensitivity* curve of the right-hand diagram in which the peak load Λ^M is plotted as a function of the imperfection parameter ϵ. This latter curve has the local form

$$\text{(load reduction)} \propto \text{(imperfection parameter)}^{1/2} \qquad (59)$$

giving us an infinite slope at Λ^C which implies a severe sensitivity for small values of ϵ. This type of bifurcation arises in the buckling of certain shells and frames.

The stable-symmetric point of bifurcation of Fig.6 perhaps requires little comment as the response is well-known in the buckling of struts and plates. In the absence of material yielding imperfect systems exhibit no failure load, the deflections simply growing rapidly as the critical load of the perfect system is approached.

The unstable-symmetric point of bifurcation of Fig.7 is quite different. Here we have an initial imperfection-sensitivity of the form

$$\text{(load reduction)} \propto \text{(imperfection parameter)}^{2/3} \qquad (60)$$

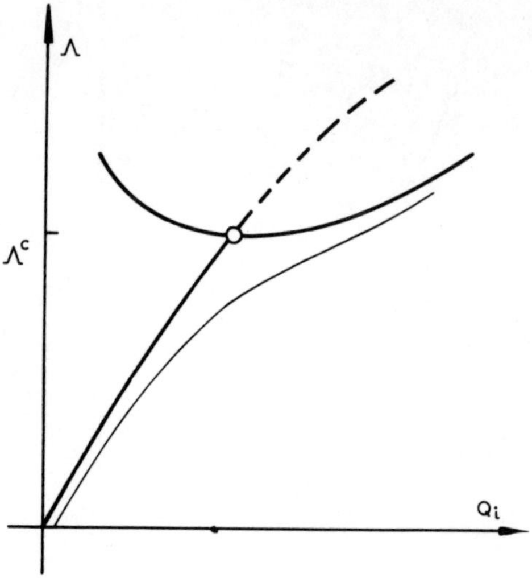

Fig. 6 The stable-symmetric point of bifurcation familiar in the buckling of struts and plates.

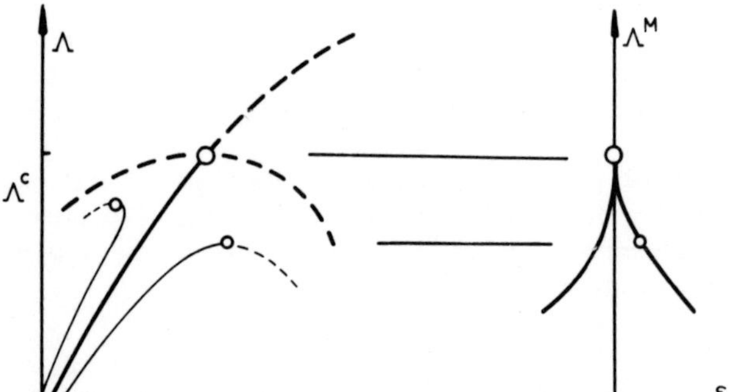

Fig. 7 The unstable-symmetric point of bifurcation exhibiting severe imperfection-sensitivity. This point of bifurcation is exhibited by a wide variety of shell-buckling problems.

with again an infinite slope at $\Lambda = \Lambda^C$ so that an extreme sensitivity to small imperfections is implied. This point of bifurcation is most common in the buckling of thin elastic shells.

9 Concluding remarks

We have here restricted attention to the distinct branching points of elastic stability, and in conclusion we must observe that many important *coupling phenomena* arise in the more complex *compound* branching points which are often generated by simple optimization schemes [1, 8–10].

References

1 Thompson, J. M. T. and Hunt, G. W. *A General Theory of Elastic Stability,* Wiley, London (1973).

2 Croll, J. G. A. and Walker, A. C. *Elements of Structural Stability*, Macmillan, London (1972).

3 Thompson, J. M. T. and Hunt, G. W. 'Comparative perturbation studies of the elastica', *Int. J. Mech. Sci.,* vol.11 (1969), pp.999–1014.

4 Thompson, J. M. T. 'Localized Rayleigh functions for structural and stress analysis', *Int. J. Solids Structures,* vol.3 (1967), pp.285–292.

5 Thompson, J. M. T. 'Basic theorems of elastic stability', *Int. J. Engng Sci.,* vol.8 (1970), pp.307–313.

6 Thompson, J. M. T. and Lewis, G. M. 'Continuum and finite-element branching studies of the circular plate', *Computers and Structures,* vol.2 (1972), pp.511–534.

7 Thompson, J. M. T. 'A general theory for the equilibrium and stability of discrete conservative systems', *Z. angew. Math. Phys.,* vol.20 (1969), p.797.

8 Thompson, J. M. T. and Lewis, G. M. 'On the optimum design of thin walled compression members', *J. Mech. Phys. Solids,* vol.20 (1972), p.101–109.

9 Thompson, J. M. T. and Supple, W. J. 'Erosion of optimum designs by compound branching phenomena', *J. Mech. Phys. Solids,* vol.21 (1973), pp.135–144.

10 Thompson, J. M. T. and Hunt, G. W. 'Dangers of structural optimization', International Symposium on Optimization in Civil Engineering, Liverpool, 16–20 July 1973. To be published in *Engineering Optimization,* a new journal edited by A. B. Templeman.

3 Coupled buckling modes of structures

W. J. Supple*

1 Introduction

Investigations of the elastic stability of structures have shown that a structure when loaded may lose its stability at a *limit point* or at a *point of bifurcation.* The former is the name given to a delimiting local maximum value of the fundamental equilibrium path in load-configuration space and the latter represents a point of intersection of this path with another distinct equilibrium path. Limit points may be established and investigated by means of a nonlinear analysis and bifurcation points may be predicted by a linear eigenvalue analysis. The term fundamental equilibrium path here refers to that path that the structure follows from the onset of loading and it may happen that there exist many (possibly an infinity of) bifurcation points on this path. Naturally, interest centres on the bifurcation point which occurs at the lowest level of load since this represents the critical or buckling load of the structure. In a loading history the load level at this point represents an upper bound to the load values for which the structure may remain in the unbuckled state. When only one other equilibrium path emanates from the fundamental path at this point (i.e. the lowest eigenvalue has a unique eigenvector) and there are no other bifurcation points on the fundamental path within a given range of load termed the *region of interest* then the point of bifurcation is said to be *distinct.* The post-buckling equilibrium path corresponding to a distinct point of bifurcation may adopt one of three distinct forms described as asymmetric, stable-symmetric and unstable-symmetric and these have been discussed by Thompson in Chapter 2 and elsewhere [1] in great detail. The forms of these paths and their importance in governing the stability behaviour of structures were first established by Koiter [2].

The profound importance of post-buckling characteristics on the stability of engineering structures lies in the simple fact that at or near the critical load there exist other adjacent equilibrium paths besides the fundamental path. Furthermore, these other equilibrium states may exist at loads less than the critical value and hence create potentially harmful properties with regard to load-carrying capability. In this respect the occurrence of initial imperfections in geometry or loading and the resulting phenomenon of *imperfection-sensitivity* must be mentioned.

With this knowledge in mind it would seem wise to be aware of *all* possible post-buckling equilibrium paths and their forms that might occur in the region of interest in any particular stability problem. We have said that by definition a

* Department of Civil Engineering, University of Surrey.

distinct point of bifurcation has only one post-buckling equilibrium path the forms of which are now well-understood. We are logically led to pose the questions when and how may other post-buckling paths be generated in the region of interest? Since post-buckling paths bifurcating from the fundamental state are of crucial importance to observed structural behaviour then the probability of paths bifurcating (say) from the post-buckling path may be of equal importance. The existence of a multiplicity of equilibrium paths at or near the critical load level has far-reaching consequences for numerical nonlinear stability analysis. Numerical methods which converge to an equilibrium solution from an approximate predicted solution using an iterative process may experience difficulty when there are many equilibrium solutions in the region of the initial guess. We may say that such situations would be ill-conditioned for numerical investigation of this type. Furthermore, if there is such a proliferation of buckling modes at the critical load then accurate instability predictions based on *initial post-buckling behaviour* which have been so successful in the case of distinct bifurcation points may be difficult and/or costly to achieve. In such instances a return to the concept of *minimum post-buckling load* which occupied the attention of early investigators of post-buckling behaviour [3] may be the only viable theoretical approach and a heavy reliance on experimental investigations may well be justified.

In the present paper we examine the conditions under which equilibrium paths through critical points are generated. We shall see that when two or more of the critical points (eigenvalues) on the fundamental equilibrium path hereafter referred to as *primary points of bifurcation* become nearly coincident, then *secondary points of bifurcation* with associated equilibrium paths appear on the post-buckling paths emanating from the primary critical points. These secondary points of bifurcation become coincident with the primary points of bifurcation when the primary points themselves become coincident. This means then that when the primary critical points coalesce on the fundamental equilibrium path (this condition is often referred to as *simultaneous buckling*) then a whole cluster of post-buckling equilibrium paths may emerge from the compound branching point thus formed.

What should be the basis to begin a study of these phenomena? We shall adopt the following procedure: when the critical load corresponds to a multiply-compound branching point we shall wish to determine the probable total number of post-buckling paths which may be generated; when the critical point is simply a doubly-compound branching point we shall examine the number and form of the possible post-buckling paths in detail and attempt to categorize the forms of behaviour of the corresponding structural systems.

2 Energy formulation

The concept of generalized coordinates in the general theory of elastic stability is well established and will be used in the following analysis without reiteration of the well-known principles involved. We shall consider a well-behaved structural system of n degrees of freedom the deflected form of which is defined completely by the n generalized coordinates Q_i $(i = 1, 2, \ldots n)$. We suppose the structural system to be subjected to a set of conservative external forces each of which is specified as a function of a single, variable load parameter Λ; non-conservative forces will not be considered. The total potential energy V of such a structural system may be written as a function of the load and deflection parameters in the form

$$V = V(\Lambda, Q_i) \tag{1}$$

where $V(\Lambda, Q_i)$ is a well-behaved single-valued function. We postulate that the partial derivatives of V with respect to the Q_i and Λ always exist and are always continuous within the region of interest. Once a value of Λ has been prescribed then this energy function represents a 'surface' in the $(n + 1)$ dimensional configuration space represented by the axes V, Q_i. We can imagine new energy surfaces for different but constant values of Λ. As was pointed out by Thompson in Chapter 2, equilibrium paths may be thought of as the loci of stationary points on these sets of energy surfaces or in other words the condition for statical equilibrium of the structural system is that the energy function $V(\Lambda, Q_i)$ is stationary with respect to all n generalized coordinates simultaneously which implies

$$V_i(\Lambda, Q_i) = 0 \tag{2}$$

where here and in the following a subscript i on energy symbols indicates partial differentiation with respect to the corresponding generalized coordinate Q_i. As i takes the range of values $1, 2, 3 \ldots n$ then equation (2) represents a set of n simultaneous equations which must be satisfied for equilibrium. Following the arguments of Chapter 2 we may suppose that one solution of these equilibrium equations $Q_i = Q_i^F(\Lambda)$ defines a fundamental state equilibrium path which is single-valued in the region of interest and passes through the origin of Λ-Q_i space; we see, therefore, that the fundamental state represents the initial loading path of the structure as Λ is increased from zero.

Further we may introduce incremental sliding coordinates q_i given by

$$Q_i = Q_i^F(\Lambda) + q_i \tag{3}$$

thus introducing Thompson's W energy function the quadratic form of the second variation of which may be diagonalized to produce the convenient $A(u_i, \Lambda)$ energy function in terms of *principal* generalized coordinates u_i. It is of interest to note that if there is symmetry in at least $(n - 1)$ of the generalized coordinates q_i such that

$$W(\Lambda, q_1, q_2, \ldots q_n) = W(\Lambda, -q_1, q_2, \ldots q_n) = W(\Lambda, q_1, -q_2, \ldots q_n) = \text{etc} \tag{4}$$

then $W_{ij}(0, \Lambda)$ is automatically diagonalized and the q_i therefore represent the principal generalized coordinates and may be written straightway in the conventional u notation. Considering systems for which W and hence A are linear in Λ we examine the stability of these systems in the vicinity of a point on the fundamental state at $\Lambda = \Lambda_0$. Suppose

$$\Lambda = \Lambda_0 + \lambda \tag{5}$$

where λ is an incremental change from the load Λ_0, then we may write the energy function $A(\Lambda, u_i)$ as

$$A(\Lambda, u_i) = A(\Lambda_0 + \lambda, u_i). \tag{6}$$

On expanding the right hand side of equation (6) in the form of a Taylor's series and employing the dummy suffix summation convention we obtain

$$\begin{aligned} A(\Lambda_0 + \lambda, u_i) = {} & A(\Lambda_0, 0) + A_i u_i + \frac{1}{2!} A_{ii} u_i^2 \\ & + \frac{1}{3!} A_{ijk} u_i u_j u_k + \dots \\ & + \lambda [A' + A'_i u_i + \frac{1}{2!} A'_{ii} u_i^2 + \frac{1}{3!} A'_{ijk} u_i u_j u_k \\ & + \dots], \end{aligned} \tag{7}$$

in which a prime denotes partial differentiation with respect to Λ and all derivatives are evaluated at the point $\{\Lambda = \Lambda_0, u_i = 0\}$ it further being understood that all subscripts take the range of values one to n.

The equations of equilibrium are now given by the conditions for stationary values of the A energy function

$$A_i(\Lambda, u_i) = 0 \tag{8}$$

which may be written in the power series forms using equation (7) as

$$\begin{aligned} A_1(\Lambda, u_i) = {} & A_{11} u_1 + \frac{1}{2!} A_{1ij} u_i u_j + \frac{1}{3!} A_{1ijk} u_i u_j u_k + \dots \\ & + \lambda \left[A'_{11} u_1 + \frac{1}{2!} A'_{1ij} u_i u_j + \dots \right] = 0, \\ & \vdots \\ A_r(\Lambda, u_i) = {} & A_{rr} u_r + \frac{1}{2!} A_{rij} u_i u_j + \frac{1}{3!} A_{rijk} u_i u_j u_k + \dots \\ & + \lambda \left[A'_{rr} u_r + \frac{1}{2!} A'_{rij} u_i u_j + \dots \right] = 0, \\ & \vdots \\ A_n(\Lambda, u_i) = {} & A_{nn} u_n + \frac{1}{2!} A_{nij} u_i u_j + \frac{1}{3!} A_{nijk} u_i u_j u_k + \dots \\ & + \lambda \left[A'_{nn} u_n + \frac{1}{2!} A'_{nij} u_i u_j + \dots \right] = 0. \end{aligned} \tag{9}$$

The coefficients A_{ii} form a set of stability coefficients sometimes written as C_i (see Chapter 2). This can be seen clearly without reference to the second variation of the energy $\delta^2 A$ by writing equations (9) in their most approximate form by neglecting higher order terms in u_i and λ as

$$
\begin{aligned}
A_{11}u_1 &= 0, \\
A_{22}u_2 &= 0, \\
&\;\vdots \\
A_{nn}u_n &= 0
\end{aligned} \tag{10}
$$

for which we see that non-trivial solutions exist if all or any of the A_{ii} coefficients are equal to zero. When $A_{rr} = 0$, that is when $\{\partial^2 A/\partial u_r^2\}_{\Lambda=\Lambda_0} = 0$ then at the load Λ_0 there is a buckling mode the form of which is initially predominant in the u_r generalized coordinate. We see then that the conditions $A_{ii} = 0$ with i ranging from one to n represent a set of *n primary bifurcation points* on the fundamental state equilibrium path. Consider the case when $A_{11} = 0$ and all other coefficients A_{ii} $(i \neq 1)$ are positive and small, approaching zero. This means that at $\Lambda = \Lambda_0$ the derivative $\partial^2 A/\partial u_1^2$ is equal to zero and all other derivatives $\{\partial^2 A/\partial u_i^2\}\,(i \neq 1)$ have small non-zero values; at values of Λ less than Λ_0 we consider all these derivatives to have positive values which implies that $A'_{ii} < 0$. We also consider that at values of Λ slightly greater than Λ_0, the derivatives A_{ii} $(i \neq 1)$ each become zero; the particular derivative A_{rr} becomes zero at the load given by

$$\Lambda = \Lambda_0 + \Delta\lambda_{(r-1)}. \tag{11}$$

We have thus defined $(n - 1)$ increments of load $\Delta\lambda_i$ $(i \neq n)$ representing the differences in load level between the lowest primary bifurcation point at Λ_0 and the other $(n - 1)$ such points on the fundamental state. Since the A_{ii} are linear functions of Λ we may write

$$A_{rr} + A'_{rr}\Delta\lambda_{(r-1)} = 0 \tag{12}$$

and remembering that $A_{11} = 0$ we can substitute from this equation for the A_{ii} into the equilibrium equations (9) to give them in the form

$$
\begin{aligned}
&\frac{1}{2}A_{1ij}u_iu_j + \frac{1}{3!}A_{1ijk}u_iu_ju_k + \ldots + \lambda\left[A'_{11}u_1 + \frac{1}{2}A'_{1ij}u_iu_j + \ldots\right] &= 0 \\
&\qquad\vdots \\
&\frac{1}{2}A_{rij}u_iu_j + \frac{1}{3!}A_{rijk}u_iu_ju_k + \ldots + \lambda\left[\frac{1}{2}A'_{rij}u_iu_j + \ldots\right] + (\lambda - \Delta\lambda_{(r-1)})A'_{rr}u_r &= 0 \\
&\qquad\vdots \\
&\frac{1}{2}A_{nij}u_iu_j + \frac{1}{3!}A_{nijk}u_iu_ju_k + \ldots + \lambda\left[\frac{1}{2}A'_{nij}u_iu_j + \ldots\right] + (\lambda - \Delta\lambda_{(n-1)})A'_{nn}u_n &= 0
\end{aligned} \tag{13}
$$

To determine initial post-buckling behaviour we may truncate these equations to any sufficient degree of approximation; for instance, if all the coefficients shown are non-zero then we may truncate systematically to give the following first order post-buckling solutions:

$$\frac{1}{2}A_{1ij}u_iu_j + \lambda A'_{11}u_1 = 0$$

$$\frac{1}{2}A_{rij}u_iu_j + (\lambda - \Delta\lambda_{(r-1)})A'_{rr}u_r = 0$$

$$\frac{1}{2}A_{nij}u_iu_j + (\lambda - \Delta\lambda_{(n-1)})A'_{nn}u_n = 0 \tag{14}$$

Any coupling between the buckling modes will be demonstrated by relationships between the generalized displacement coordinates, a fact that will be discussed in detail in the sequel. We may obtain such relationships by eliminating λ between equations (14) to give the following $(n-1)$ equations:

$$\frac{1}{2}(A_{2ij}A'_{11}u_1 - A_{1ij}A'_{22}u_2)u_iu_j = A'_{11}A'_{22}u_1u_2\Delta\lambda_1$$

$$\frac{1}{2}(A_{rij}A'_{11}u_1 - A_{1ij}A'_{rr}u_r)u_iu_j = A'_{11}A'_{rr}u_1u_r\Delta\lambda_{(r-1)}$$

$$\frac{1}{2}(A_{nij}A'_{11}u_1 - A_{1ij}A'_{nn}u_n)u_iu_j = A'_{11}A'_{nn}u_1u_n\Delta\lambda_{(n-1)}. \tag{15}$$

We may use the equations derived here to establish conditions for the existence of equilibrium paths in the region of interest. The occurrence of secondary, tertiary or higher points of bifurcation as indicated in the introduction appears possible. The most general situation is when all the primary bifurcation points are non-coincident but sufficiently close together so as not to warrant the description *distinct*. Coincidence of some or indeed all of the primary bifurcation points may then be viewed as special cases of the non-coincident non-distinct situation. It has been shown by Thompson and Supple [4] that a close examination of the post-buckling characteristics during the transition from the latter case to the coincident case is very important when stability-based optimization procedures are performed.

3 Multiple post-buckling paths

Non-trivial solutions of equations (14) will represent post-buckling equilibrium solutions. We may define any solution, if it exists, that involves only one of the u_i

displacement parameters as an *uncoupled* solution and any solution that involves more than one of u_i as a *coupled* solution. This terminology is self-explanatory and will be found to be useful in the ensuing discussion. Let us assume that coupled solutions may occur and if so calculate how many such modes are possible. First let us assume that coupling between r of the n displacement parameters occurs but for the time being make no statement about the number of equilibrium paths corresponding to a specific coupling of a particular set of r parameters. The total number N_r of such couplings is

$$N_r = {}^nC_r \tag{16}$$

where ${}^nC_r$ represents the number of ways of selecting r items from n items. We note that when $r = 1$ the possible solutions are the n uncoupled solutions $N_1 = n$. The total number then of all combinations of deformation coordinates which may form buckling modes is

$$N = \sum_{r=1}^{n} N_r = \sum_{r=1}^{n} {}^nC_r \tag{17}$$

but

$$\sum_{r=1}^{n} {}^nC_r = 2^n - 1 \tag{18}$$

so we see that the total number of ways of selecting combinations of the u_i is given by

$$N = 2^n - 1\,. \tag{19}$$

We may argue that each of such combinations may result in more than one post-buckling equilibrium path, so that the total number of such paths is still undetermined. We now turn our attention to the work of Chilver and Johns [5] who have investigated in some detail the coincident buckling situation. These authors have shown that the total number of post-buckling equilibrium paths may increase beyond the number N given by equation (19) if sufficient symmetry is introduced into the system. This increase in the number of possible post-buckling equilibrium paths due to symmetry had been noted earlier for two-degree-of-freedom systems by Chilver [6] and Supple [7].

We may summarize the results of Chilver and Johns as follows.

If we have a system with n *coincident* primary bifurcation points at the critical load and the total potential energy of the structural system is symmetric in r of the u_i coordinates and asymmetric in m of these coordinates ($r + m = n$) then when $r = 0$ or $r \leqslant m$, the system has a minimum of one and a maximum of $2^n - 1$ post-buckling paths. These paths are all represented by a linear load-coordinate relationship.

When $r > m$ the system has a minimum of one linear post-buckling path and a maximum number of paths given by

$$N = 2^m - 1 + 2^m \sum_{j=1}^{m} {}^rC_j + \sum_{j=m+1}^{r} 2^{j-1}\, {}^rC_j \tag{20}$$

in which the last term represents the maximum number of paths represented by a quadratic relationship between the load and the symmetric coordinates.

When $m = 0$ and the total potential energy of the system is symmetric in all the u_i, the minimum number of post-buckling paths is n and the maximum number is

$$N = \tfrac{1}{2}(3^n - 1). \tag{21}$$

These simple formulae illustrate vividly the potential post-buckling complexity of structures with even a small number of degrees of kinematic freedom. The following table gives the possible number of paths for values of n up to seven.

Degrees of freedom, n		1	2	3	4	5	6	7
Maximum number of post-buckling paths	From eqn (19) $N = 2^n - 1$	1	3	7	15	31	63	127
	From eqn (21) $N = \frac{1}{2}(3^n - 1)$	1	4	13	40	121	364	1093

No theorems have as yet been proposed regarding the existence of these paths in general. In the following section we shall discuss the detailed form of the coupled buckling modes for systems with two degrees of freedom – from the table above we anticipate a maximum of three or four post-buckling equilibrium paths depending upon the degree of symmetry involved.

4 Systems with two degrees of freedom

A study of systems with only two degrees of freedom is useful in that it enables us to establish the mechanics of coupling phenomena in their simplest form. Furthermore, there are systems which have several degrees of kinematic freedom but for which only two of the degrees of freedom are of interest with regards initial post-buckling behaviour. For an approximate analysis of the early stages of post-buckling we may consider such systems essentially as two-degrees-of-freedom systems. It may be pointed out that certain deformations may be justifiably ignored in a buckling analysis. To elaborate, we may distinguish between two types of deformation of a loaded structure which we term *buckling deflections* and *non-buckling deflections*. The latter, which are normally less spectacular than the buckling deflections, often constitute the fundamental equilibrium state mentioned previously; the buckling deflections are then generated by bifurcations from this fundamental state. Examples of non-buckling deflections would be the axial shortening of an Euler strut, or the end shortening of a rectangular plate or circular cylindrical shell under compression between the ends, or again, the radial contraction of a spherical shell under uniform external pressure. For an approximate but reasonably exact analysis of these structures we may ignore the non-buckling deflections. The deformation parameters will then represent only buckling deflections and the fundamental state becomes the trivial path coincident with the load axis in Λ-Q_i space.

When $n = 2$ equations (14) become

$$\tfrac{1}{2}(A_{111}u_1^2 + 2A_{112}u_1u_2 + A_{122}u_2^2) + \lambda A'_{11}u_1 = 0$$

$$\tfrac{1}{2}(A_{112}u_1^2 + 2A_{122}u_1u_2 + A_{222}u_2^2) + (\lambda - \Delta\lambda_1)A'_{22}u_2 = 0 \qquad (22)$$

and the u_1-u_2 relationship represented by equations (15) for the general system reduces to the following single equation for the two-coordinate system:

$$A_{112}A'_{11}u_1^3 + (2A_{122}A'_{11} - A_{111}A'_{22})u_1^2u_2 + (A_{222}A'_{11} - 2A_{112}A'_{22})u_1u_2^2$$
$$- A_{122}A'_{22}u_2^3 = 2A'_{11}A'_{22}u_1u_2\Delta\lambda_1. \qquad (23)$$

Since there are only two primary bifurcation points to be considered we may drop the subscript from $\Delta\lambda_1$ and write it simply as $\Delta\lambda$ in the continuing analysis.

In the vicinity of the lower critical point (i.e. lower primary bifurcation point) u_1, u_2 and λ may be considered as being small compared with $\Delta\lambda$ and equations (22) may be further approximated to give the asymptotically correct solution

$$\lambda = -\tfrac{1}{2}\frac{A_{111}}{A'_{11}}u_1, \quad u_2 = 0 \qquad (24)$$

which represents a linear equilibrium path in the u_1 mode.

Similarly, in the vicinity of the upper critical point u_1, u_2 and $(\lambda - \Delta\lambda)$ may be considered as being small compared with λ with the resulting solution from equations (22) as

$$\lambda = \Delta\lambda - \frac{A_{222}}{2A'_{22}}u_2, \quad u_1 = 0 \tag{25}$$

which represents a linear equilibrium path in the u_2 mode.

In the non-coincident non-distinct situation, therefore, we have branching from the fundamental state at the two primary bifurcation points in the u_1 and u_2 directions as indicated by equations (24) and (25). When we have coincidence of the critical loads then the arguments used in the derivation of equations (24) and (25) are no longer valid so we now focus our attention on equation (23) the right-hand side of which is now zero since $\Delta\lambda = 0$. The resulting cubic in u_1 and u_2 may have one or three real roots and may be expressed in the form

$$(u_1 + \bar{a}u_2)(u_1 + \bar{\beta}u_2)(u_1 + \bar{\gamma}u_2) = 0 \tag{26}$$

in which the coefficients $\bar{a}$, $\bar{\beta}$, $\bar{\gamma}$ are functions of the A coefficients in equation (23). We see from equation (26) that the post-buckling solutions when real are straight lines in the u_1-u_2 projection. For larger values of u_1 and u_2 the equilibrium paths for the non-coincident case must approach those for the simultaneous buckling case by reason of the fact that the right-hand side of equation (23) becomes relatively small and hence negligible under these conditions and so equation (26) governs.

The qualitative form of the single and triple root solutions is given in Fig.1 for the separated and compound buckling cases as represented by their u_1-u_2 projections. These solutions and the equations from which they were derived were first proposed by Chilver [6]. As $\Delta\lambda \rightarrow 0$ then the separated forms degenerate to the linear coincident forms. A simple conceptual aid in envisaging this process is to think of the equilibrium paths as strings which are gradually pulled taut; in the single-path case we see then that the 'loop' is gradually eliminated. These results indicate clearly that if we have near-coincident buckling conditions then violent path contortions are possible in the initial post-buckling regime.

We note in passing that the maximum number of three post-buckling paths predicted by equation (19) may be attained here provided the energy coefficients are such that three real roots of equation (26) exist.

5 Symmetrical systems

We shall now consider systems with two degrees of freedom and whose total potential energy function is an even function of either one or both of the displacement coordinates u_1 and u_2. That is, we define two broad classes of structural systems for which either

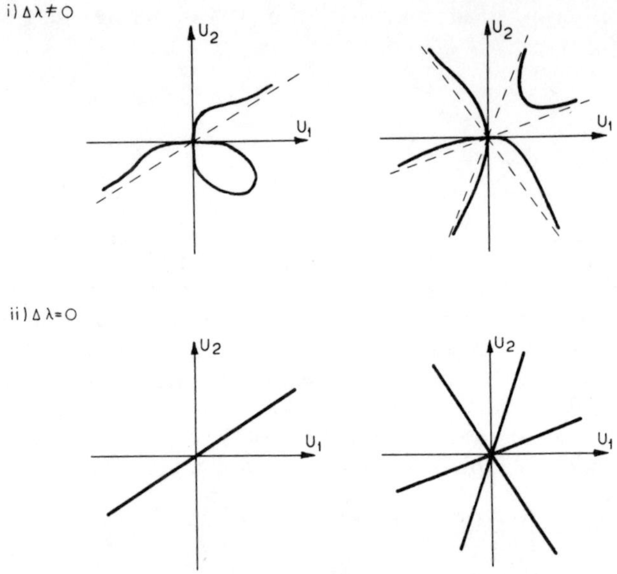

Fig. 1 Forms of u_1-u_2 relationship for general asymmetric system.

$$\text{(i)}\quad A(\Lambda, u_1, u_2) = A(\Lambda, u_1, -u_2) \tag{27}$$

and which we shall refer to as singly-symmetric systems or

$$\text{(ii)}\quad A(\Lambda, u_1, u_2) = A(\Lambda, u_1, -u_2) = A(\Lambda, -u_1, u_2) \tag{28}$$

which we shall refer to as doubly-symmetric systems. Engineering structures often display symmetry properties of the forms given by equations (27) and (28) arising from the nature of the design process which may employ symmetrical manufactured components and which aims at aesthetic balance. Therefore, imposition of conditions (27) and (28) does not imply a great loss of generality in the analysis.

If we now turn our attention to singly-symmetric systems we see that the equilibrium equations (22) reduce to the forms

$$\left.\begin{aligned} \tfrac{1}{2}(A_{111}u_1^2 + A_{122}u_2^2) + \lambda A'_{11}u_1 &= 0 \\ A_{122}u_1u_2 + (\lambda - \Delta\lambda)A'_{22}u_2 &= 0 \end{aligned}\right\} \tag{29}$$

where the coefficients A_{112}, A_{222} have disappeared owing to the symmetry in the u_2 parameter. Also, the u_1-u_2 relationship given by equation (23) is likewise simplified to appear now as

$$K_1u_1 + K_2u_1^2 + K_3u_2^2 = 0 \tag{30}$$

where

$$K_1 = 2A'_{11}A'_{22}\Delta\lambda$$

$$K_2 = (A_{111}A'_{22} - 2A_{122}A'_{11})$$

$$K_3 = A_{122}A'_{22}\ .$$

We may distinguish two post-buckling solutions of equations (29) which are characterized by

$$\left.\begin{array}{l}\text{(i) } u_1 \neq 0, \quad u_2 = 0\\[2ex] \text{(ii) } u_1 \neq 0, \quad u_2 \neq 0\end{array}\right\} \tag{31}$$

which we find by virtue of our earlier definition may be termed an uncoupled and coupled buckling mode respectively. Solution (31 (i)) may be written in full as

$$\lambda = -\frac{A_{111}}{2A'_{11}}u_1, \quad u_2 = 0 \tag{32}$$

which is typical of the asymmetric post-buckling bifurcation at a distinct bifurcation point, see Chapter 2, and is illustrated in Fig.2. When we examine solution (31(ii)) we discover that the λ-u_1 relationship is given directly by the second of equations (29) and may be written as

$$\lambda = \Delta\lambda - \frac{A_{122}}{A'_{22}}u_1 \tag{33}$$

which is a straight line (indicating that the coupled mode lies in a plane) and therefore, in general, intersects the uncoupled equilibrium path in one point (Fig.2). This point corresponds to a point of bifurcation at which the coupled mode branches from the uncoupled post-buckling equilibrium path. The straight line given by equation (33), representing the projection of the coupled mode on the λ-u_1 plane, crosses the load axis at the value $\Lambda_0 + \Delta\lambda$; the coupled mode at this point corresponds to the second critical branching solution on the fundamental state. It is apparent that only one secondary point of bifurcation has been generated.

The asymptotes of the family of curves represented by equation (30) are given by

$$u_2 = \pm\left(-\frac{K_2}{K_1}\right)u_1 \mp \frac{K_1}{2(-K_2K_3)^{1/2}} \tag{34}$$

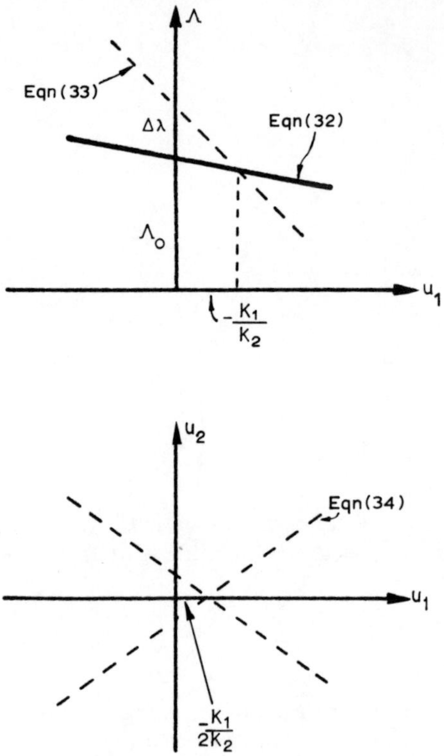

Fig. 2 Uncoupled post-buckling equilibrium path and asymptotes of coupled solution for singly-symmetric system.

and are also illustrated in Fig.2. These asymptotes are real only when the product $-K_2K_3$ is positive. Equation (30) may adopt the forms of an ellipse or a hyperbola branching from the u_1 axis (Fig.3). When the two critical points on the fundamental state coincide (simultaneous buckling condition $\Delta\lambda = 0$) then the ellipse solution collapses to a point at the origin and the hyperbola solution degenerates to its asymptotes through the origin. The spatial form of the equilibrium paths may be indicated by and deduced from their projections onto the coordinate planes and these are depicted for the hyperbola-type coupled solution for the three cases $\Delta\lambda > 0$, $\Delta\lambda = 0$, $\Delta\lambda < 0$ in Fig.4. The most important point to note from this figure is that the branching solution associated with the secondary point of bifurcation is characteristic of the unstable-symmetric post-buckling encountered at distinct primary critical points, i.e. this is a 'falling' path. The introduction of imperfection sensitivity associated with this falling path is an obvious danger. Further, if the plane of the coupled mode slopes in the opposite direction to that

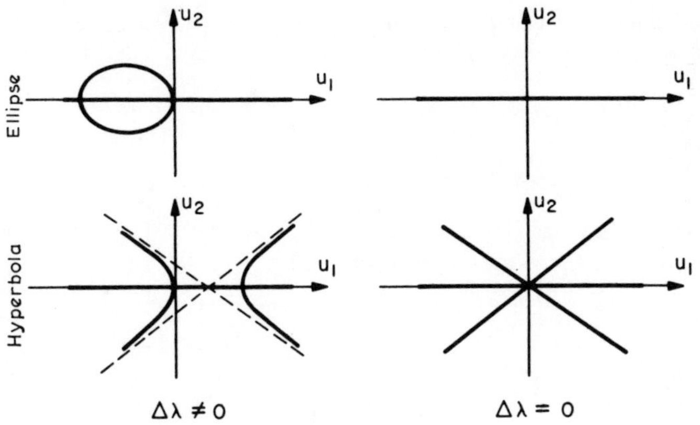

Fig. 3 Forms of u_1-u_2 relationship for coupled mode for singly-symmetric system. system.

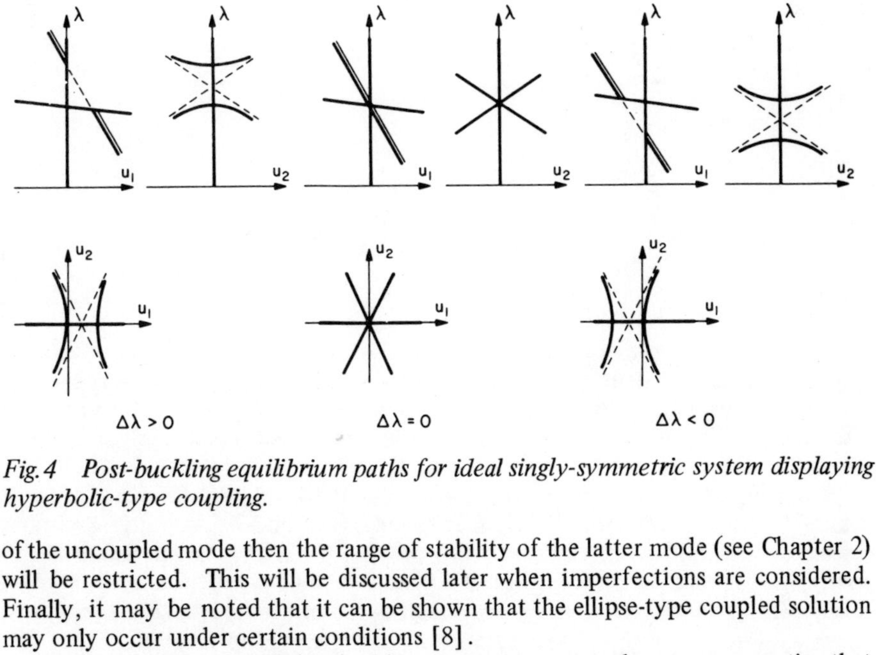

Fig. 4 Post-buckling equilibrium paths for ideal singly-symmetric system displaying hyperbolic-type coupling.

of the uncoupled mode then the range of stability of the latter mode (see Chapter 2) will be restricted. This will be discussed later when imperfections are considered. Finally, it may be noted that it can be shown that the ellipse-type coupled solution may only occur under certain conditions [8].

If we now wish to consider doubly-symmetric structural systems we notice that equations (22) are insufficient for our purposes because the series truncations implicit in these equations are too severe now that most of the lower-order energy coefficients in the series have vanished due to symmetry. In consequence, we merely return to equations (13) and truncate these in the light of the present circumstances and retain sufficient terms in the power series to establish initial post-buckling behaviour. The equilibrium equations for the doubly-symmetric system with two degrees of freedom then appear as

$$\tfrac{1}{6}(A_{1111}u_1^3 + 3A_{1122}u_1u_2^2) + \lambda A'_{11}u_1 = 0$$

$$\tfrac{1}{6}(A_{2222}u_2^3 + 3A_{1122}u_1^2u_2) + (\lambda - \Delta\lambda)A'_{22}u_2 = 0 . \qquad (35)$$

By inspection we see that besides the trivial solution $u_1 = u_2 = 0$ there are three possible solutions to these equations described by

(i) $u_1 \neq 0, \quad u_2 = 0$

(ii) $u_1 = 0, \quad u_2 \neq 0$

(iii) $u_1 \neq 0, \quad u_2 \neq 0$

which by our previous definitions are two uncoupled and one coupled mode respectively. Writing the uncoupled modes in explicit form:

$$\lambda = -\tfrac{1}{6}\,\frac{A_{1111}}{A'_{11}}\,u_1^2, \quad u_2 = 0 \qquad (37)$$

and

$$\lambda = \Delta\lambda - \tfrac{1}{6}\,\frac{A_{2222}}{A'_{22}}\,u_2^2, \quad u_1 = 0 \qquad (38)$$

we see that they represent either stable-symmetric or unstable-symmetric modes dependent upon the signs of the coefficients A_{1111} and A_{2222} (note: A'_{11} and A'_{22} are both negative by prior definition). There are four possible arrangements of these modes for any doubly-symmetric system and these are illustrated in three-dimensional plot in Fig.5.

In our arguments so far we have spoken about the primary bifurcation points coming together or coalescing to produce simultaneous buckling conditions without explaining in detail the mechanics of this process; the following comments on this point are in order. There is in general a parameter β of a structural system which governs the relative sizes of the energy coefficients in the foregoing analysis. In a practical system β might appear as a geometric relationship or as a ratio of two spring constants or material constants, etc. In an optimum design process based on simultaneous mode design β would be the variable design parameter. We postulate that by giving β a slight variation the small quantity A_{22}/A'_{22}, and hence $\Delta\lambda$, may change in magnitude and indeed may even change sign without any significant change in any of the other energy coefficients. It can be seen that if β is varied such that $\Delta\lambda$ changes from a positive value to a negative value the general forms of

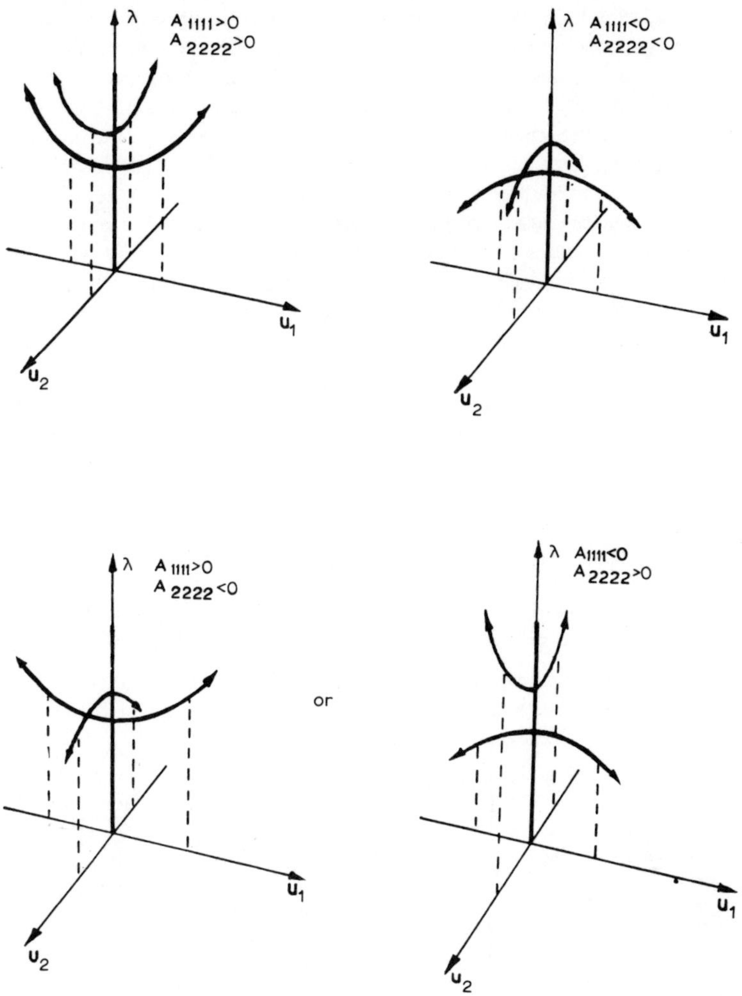

Fig.5 Uncoupled branching configurations for doubly-symmetric structural systems.

the first two systems depicted in Fig.5 do not alter and the forms of the two last systems merely interchange. In terms of a varying β, therefore, Fig.5 represents in essence not four but three systems.

Let us now consider the coupled mode (36(iii)); the u_1-u_2 relationship for this mode may be obtained in a straightforward manner by elimination of λ between equations (36) to give

$$(A'_{22}A_{1111} - 3A'_{11}A_{1122})u_1^2 + (3A'_{22}A_{1122} - A'_{11}A_{2222})u_2^2 = -6A'_{11}A'_{22}\Delta\lambda, \quad (39)$$

real solutions of which may be either an ellipse or hyperbola. Points where the ellipse or hyperbola intersect the coordinate axes represent secondary bifurcation points and so the ellipse-type coupled mode represents a *transition path* between the uncoupled modes. The forms of the coupled mode for the situation where the uncoupled modes are rising paths are illustrated in Fig.6. The mode couplings represented by the first and third cases depicted, i.e. ellipse and hyperbola branching from the lower uncoupled mode, assume immediate importance. This is because in these instances the lower coupled mode loses its stability at a secondary point of bifurcation; an analysis which ignored coupling effects would have fallaciously predicted stable post-buckling behaviour. Of all the coupling phenomena shown in Fig.6 the third type is potentially the most dangerous since there are no stable equilibrium states at loads greater than the load level at the secondary bifurcation point.

With the concept of rising and falling paths as used above Supple [7] has established the following theorems for doubly-symmetric structural systems with two degrees of freedom:

Theorem 1. When both uncoupled equilibrium paths are rising, the coupled equilibrium paths branching from the lower uncoupled path will be rising provided $A_{1111}A_{2222} > (3A_{1122})^2$ and falling provided $A_{1111}A_{2222} < (3A_{1122})^2$.

Theorem 2. When one or both uncoupled equilibrium paths are falling, the coupled paths branching from the lower uncoupled path will always be falling.

Theorem 3. When one or both uncoupled equilibrium paths are rising, the coupled paths branching from the upper uncoupled path will always be rising.

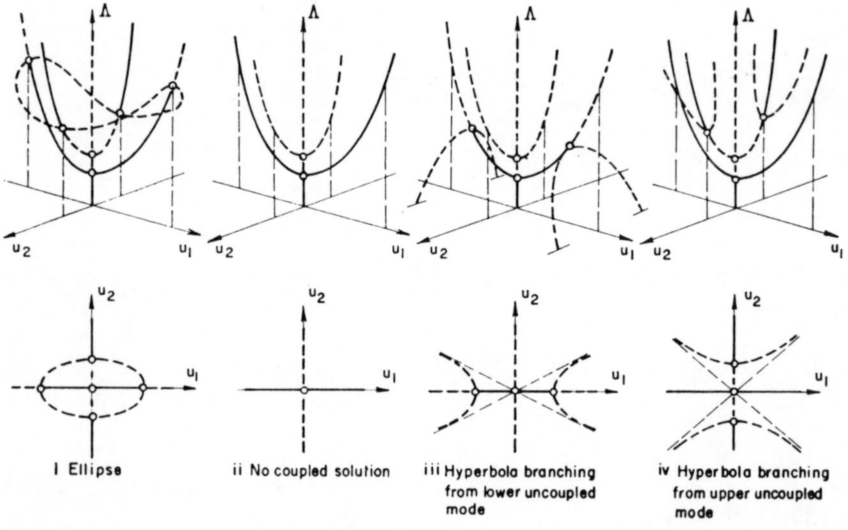

Fig.6 Forms of coupled post-buckling for ideal doubly-symmetric structural systems (——, stable equilibrium path; – – – –, unstable equilibrium path).

Theorem 4. When both uncoupled equilibrium paths are falling, the coupled path branching from the upper uncoupled path will be rising provided $A_{1111}A_{2222} < (3A_{1122})^2$ and falling provided $A_{1111}A_{2222} > (3A_{1122})^2$.

Theorem 5. If real coupled buckling modes exist when simultaneous buckling conditions prevail, then the initial curvature of either branch of the coupled post-buckling equilibrium paths will be algebraically less than the initial curvature of either uncoupled post-buckling equilibrium path provided $V_{1111} > 3V_{1122}$ and $V_{2222} > 3V_{1122}$.

6 Systems with initial imperfections

Any manufactured structural system will, in a real world, inevitably possess deviations from the desired geometrical shape, and there will always be some misalignment however small of the desired directions of the applied loads. These are termed initial imperfections. For distinct primary critical points such imperfections have been shown to have a profound effect upon the initial buckling process and have led to the concept of imperfection-sensitivity as outlined in Chapter 2. For systems exhibiting multiply-compound branching behaviour the introduction of a general imperfection would obviously lead to very complex forms of behaviour. A natural loading path would exist (on which there might be bifurcation points) and possibly a myriad of complementary loading paths would be generated which would obscure an analytical study aimed at categorizing behaviour patterns. Such an analysis would not appear at the present time to be a very fruitful area of study. However, some indication of the combined effects of coupled modes plus imperfections may be afforded by a study of the doubly-compound branching system. In the following we do just that and, further, restrict our attention to symmetrical systems. We specify that an initial imperfection ϵ exists and is such that the total potential energy which is now also a function of ϵ has the properties

$$A(\Lambda, u_1, u_2, \epsilon) = A(\Lambda, u_1, -u_2, \epsilon) \tag{40}$$

for the singly symmetric system and

$$A(\Lambda, u_1, u_2, \epsilon) = A(\Lambda, -u_1, u_2, -\epsilon) = A(\Lambda, u_1, -u_2, \epsilon) \tag{41}$$

for the doubly-symmetric system. Close examination of equations (40) and (41) reveals that we have related the imperfection ϵ to the u_1 deformation parameter in some way; in the terminology of Roorda [9] ϵ is a *major* imperfection in the u_1 mode. The effects of combined imperfections, that is, major imperfections in both the u_1 and u_2 modes have been reported elsewhere [8, 10].

Let us first consider the effect of ϵ on the singly-symmetric systems. Expanding the new energy function with ϵ included after the manner outlined by equation (7) and then differentiating with respect to u_1 and u_2 separately produces the following equilibrium equations after systematic truncation of higher-order terms in the Taylor series

$$A_{1\epsilon}\epsilon + \tfrac{1}{2}(A_{111}u_1^2 + A_{122}u_2^2) + \lambda A'_{11}u_1 = 0$$

$$A_{122}u_1u_2 + (\lambda - \Delta\lambda)A'_{22}u_2 = 0 \tag{42}$$

in which a subscript ϵ on A represents partial differentiation with respect to ϵ. We note that the only difference between equations (42) and equations (29) representing the perfect case is the inclusion of the single term $A_{1\epsilon}\epsilon$ in the first equation.

As for the perfect structural system we distinguish two solutions of equations (42), one being uncoupled, the other coupled. The uncoupled solution is

$$A_{1\epsilon}\epsilon + \tfrac{1}{2}A_{111}u_1^2 + \lambda A'_{11}u_1 = 0, \quad u_2 = 0 \tag{43}$$

which represents a family of imperfect solutions corresponding to various values of ϵ. This form of imperfect solution is well-known in connection with asymmetric branching at a distinct point of bifurcation (see Chapter 2) but for completeness is illustrated here as Fig.7. In practical structures interest centres on the natural loading paths, and in particular upon any limit points or points of bifurcation which may occur on these paths. The locus of the limit points, which represents a

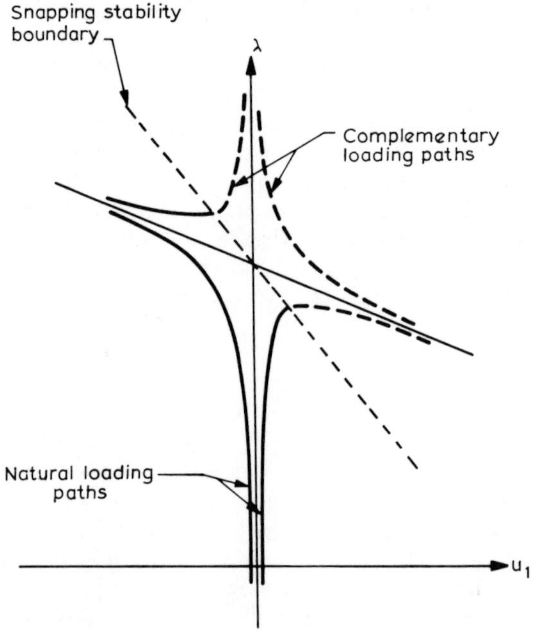

Fig. 7 Imperfect uncoupled equilibrium configurations for singly-symmetric systems.

stability boundary with regards snapping, may be obtained from equation (43) by putting $\partial\lambda/\partial u_1 = 0$ whereupon

$$\lambda = -\frac{A_{111}}{A'_{11}} u_1. \tag{44}$$

Thus, the slope of the limit point stability boundary is twice the slope of the perfect uncoupled post-buckling equilibrium path (see Fig.7).

We may now turn our attention to the imperfect coupled mode. From the second of equations (42) it is seen that this imperfect solution lies in the same plane as the perfect coupled solution; the λ-u_1 relationship is therefore given by equation (33). The u_1-u_2 relationship for the imperfect coupled mode is obtained by elimination of λ between the two equations (42) and appears in the form

$$K_1 u_1 + K_2 u_1^2 + K_3 u_2^2 + K_4 = 0 \tag{45}$$

where K_1, K_2, K_3 are as defined for equation (30) and

$$K_4 = 2A'_{22} A_{1\epsilon}\epsilon \,. \tag{46}$$

Real solutions of equation (45) represent an ellipse or hyperbola. If either the natural or complementary uncoupled loading paths or both intersect the line represented by equation (33) then branching of the imperfect uncoupled mode occurs at the points of intersection. These statements are clarified for the hyperbola-type solution and for various values of ϵ in Fig.8.

Equation (33) forms a stability boundary with respect to bifurcation into a coupled mode. In the imperfect structural systems under investigation, therefore, there exist *two stability boundaries,* one pertaining to snapping, the other to bifurcation. For various values of the imperfection parameter ϵ the natural loading path may intersect one, both or neither of the stability boundaries. When the second is the case, the stability boundary which intersects the natural path first in its loading history assumes importance. If we denote the load and deformation at which a limit point occurs as λ_L and u_L, then from equations (43) and (44) we find

$$\lambda_L^2 = 2A_{111} A_{1\epsilon}\epsilon/(A'_{11})^2 \tag{47}$$

Also, if we denote the load and deformation at which branching into a coupled mode of buckling occurs as λ_B and u_B then from equations (33) and (43) we find

$$(A_{111}A'_{22} - 2A_{122}A'_{11})\lambda_B^2 - 2\Delta\lambda(A_{111}A'_{22} - A_{122}A'_{11})\lambda_B + A_{111}A'_{22}\Delta\lambda^2 + \frac{2A_{122}^2 A_{1\epsilon}\epsilon}{A'_{22}} = 0 \tag{48}$$

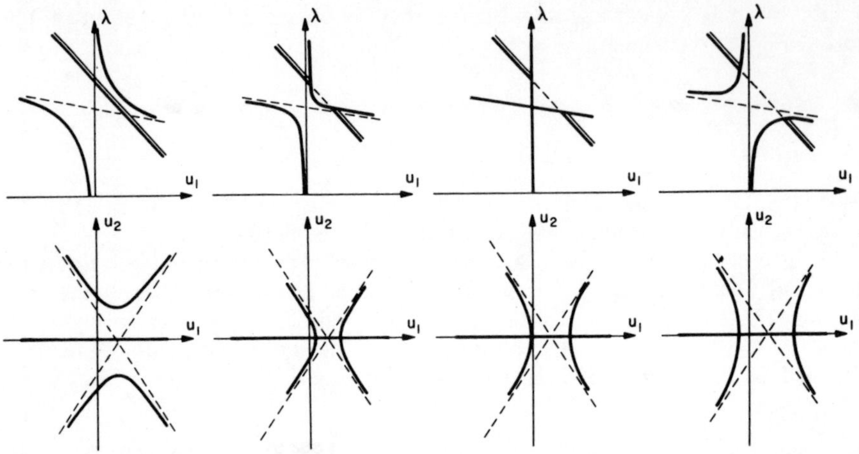

Fig. 8 Post-buckling equilibrium paths for imperfect singly-symmetric system displaying hyperbolic-type coupling, for various values of ε.

which when $\Delta\lambda = 0$ reduces to

$$\lambda_B^2 = \frac{2A_{122}^2 A_{1\epsilon}\epsilon}{A'_{22}(A_{111}A'_{22} - 2A_{122}A'_{11})} \tag{49}$$

Equations (47) and (48) establish the critical loads of the structural system for any value of the imperfection parameter ϵ; the two parabolas represented by these equations always meet at a common tangent. Similar relationships between imperfections and critical deformations may be derived from the foregoing equations. Figs 9 and 10 illustrate for $\Delta\lambda = 0$ typical loading paths and the relationships between critical loads, deformations and imperfections for structural systems displaying hyperbolic forms of coupling. Full and broken lines indicate stable and unstable equilibrium paths respectively. Fig.9 depicts the interesting case when for positive imperfections bifurcation into a coupled mode from the natural path occurs before a limit point is reached. It can be said that for structures exhibiting this type of buckling behaviour, the effect of a secondary primary branching point becoming coincident with the lowest critical point on the fundamental state is to bring about a greater degree of imperfection sensitivity. For the structural system whose behaviour is described by Fig.10 it is seen that positive imperfections lead to snap-buckling and negative imperfections lead to bifurcation buckling.

We again focus our interest on doubly-symmetric systems but now with imperfections present. The equilibrium equations are now

$$A_{1\epsilon}\epsilon + \tfrac{1}{6}(A_{1111}u_1^3 + 3A_{1122}u_1u_2^2) + \lambda A'_{11}u_1 = 0$$

$$\tfrac{1}{6}(A_{2222}u_2^3 + 3A_{1122}u_1^2u_2) + (\lambda - \Delta\lambda)A'_{22}u_2 = 0 \tag{50}$$

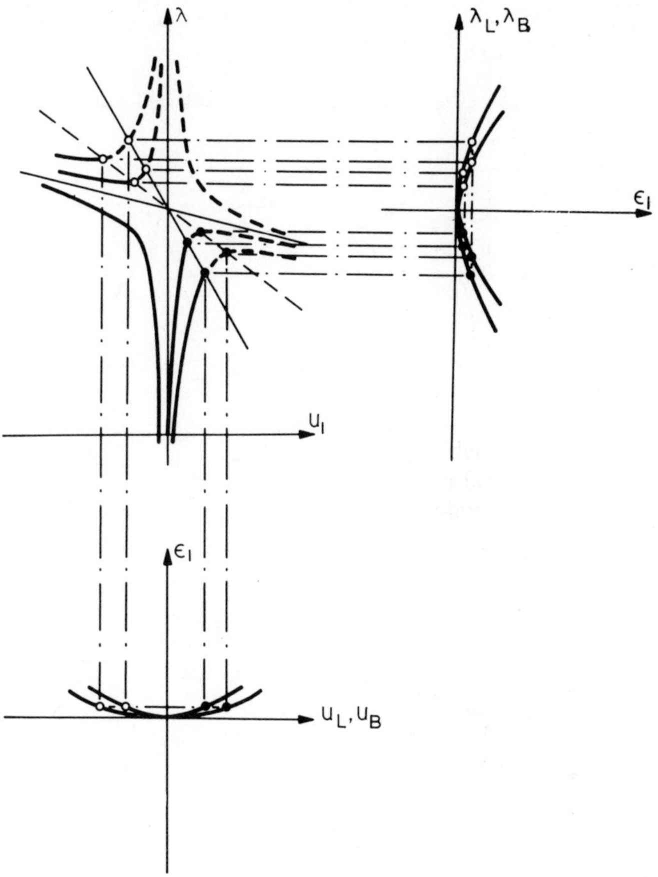

Fig. 9 Natural loading paths, critical loads and critical deformations.

where again we see that only one significant imperfection term appears. There are two solutions of equations (50), one uncoupled, the other coupled. As for the singly-symmetric case the uncoupled mode is exactly similar in form to the imperfect solutions about a distinct bifurcation point. The forms of the coupled mode become relatively complex and have been described elsewhere [10] with the general conclusion that for this type of imperfection if the primary critical points are not coincident the initial post-buckling behaviour is not drastically changed. These forms are indicated in Fig.11. However, simultaneous buckling brings about conditions for higher imperfection-sensitivity and will be outlined here briefly.

If we consider that the perfect uncoupled u_1 mode is unstable symmetric then imperfections ϵ produce limit points on the natural loading paths the locus of which is given by

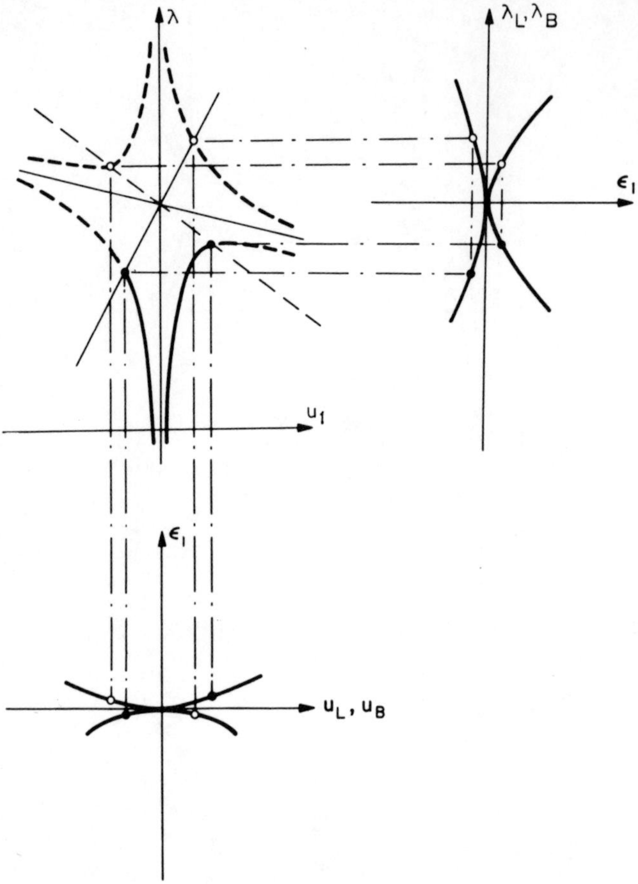

Fig. 10 Natural loading paths, critical loads and critical deformations.

$$\lambda = -\frac{A_{1111}}{2A'_{11}} u_1^2, \tag{51}$$

that is, the curvature of this limit point stability boundary is three times as great as the curvature of the perfect uncoupled mode, see equation (37). There is also a stability boundary governing bifurcation into a coupled mode the equation of which is [11]

$$\lambda = -\frac{A_{1122}}{2A'_{22}} u_1^2. \tag{52}$$

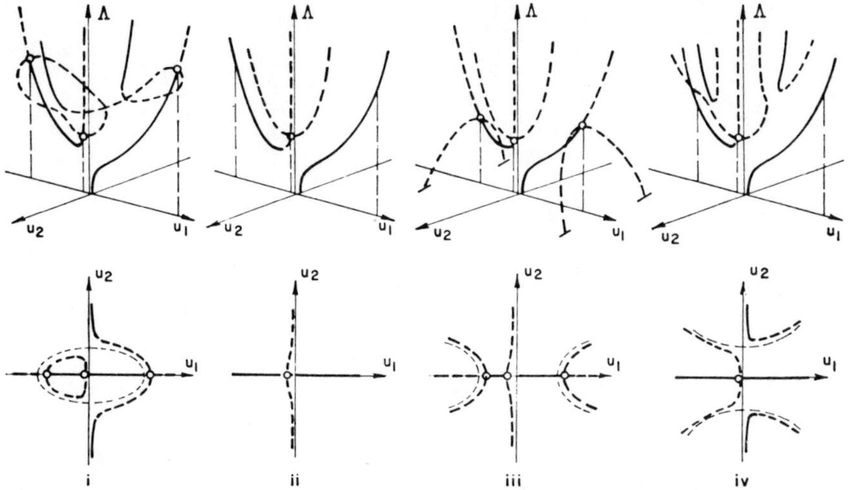

Fig. 11 Forms of post-buckling equilibrium path for doubly-symmetric structural systems with imperfections ϵ (——, stable equilibrium path; – – – –, unstable equilibrium path).

If the natural loading path intersects the latter stability boundary before intersecting that given by equation (51) then the imperfect-sensitivity is increased by compound branching. The imperfection-sensitivity equation for bifurcation is similar in form to that for snapping, that is, a two-thirds power law and may be written as

$$\lambda_B = C\epsilon^{2/3} \tag{53}$$

where

$$C = \left\{\frac{3A_{1\epsilon}A_{1122}}{A_{1111}A'_{22} - 3A'_{11}A_{1122}}\right\}^{2/3} \left\{-\frac{A_{1122}}{2A'_{22}}\right\}^{1/3} . \tag{54}$$

The situation in which bifurcation into a coupled mode occurs *before* a limit point is reached is depicted in Fig.12 where it has sufficed to show the λ-u_1 projection alone from which the results given by equations (53) and (54) were derived.

7 Conclusions

It has been shown in the previous sections that the coincidence of critical loads of structures may lead to highly complex instability characteristics. Such coincidence may arise as a natural consequence of the structure geometry and type of loading or may be the result of an optimal design process. The results of the analysis for systems with two degrees of freedom have been conveniently represented in a pictorial manner by means of three-dimensional plots of generalized load parameter

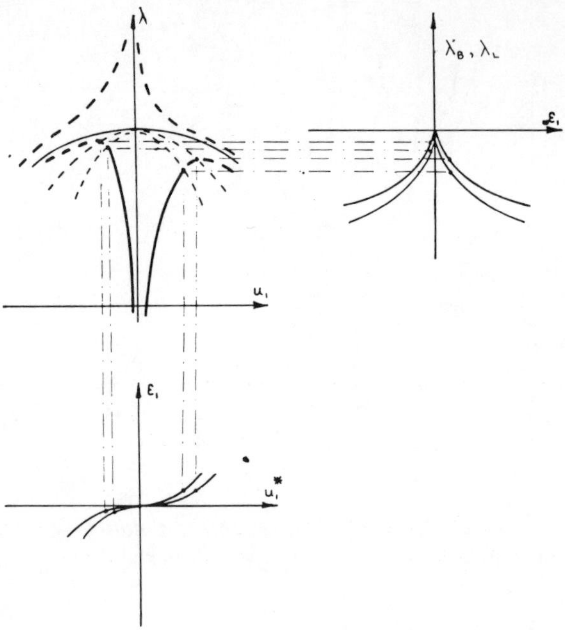

Fig. 12 Equilibrium paths and imperfection-sensitivity diagrams.

against buckling deflections. In this way the post-buckling equilibrium paths are seen to adopt simple and easily distinguished forms which have been categorized. The post-buckling complexity arises from:

(i) extreme contortion of equilibrium paths,
(ii) generation of multiple equilibrium paths.

These two factors may pose difficulties where numerical methods of solution to stability problems are employed and so the numerical analyst should be well aware of these phenomena. It would appear that many structure types belong to the class of problem defined by coincident or near-coincident buckling and do not have well separated critical loads as does the simple compressed strut. Even the latter can fall into this category if it is supported along its length by an elastic medium. Steps may be taken to separate the critical loads of a structure and improve its stability behaviour as for instance by the longitudinal stiffening or internal pressurization of a circular cylindrical shell under axial compression.

Associated with the increased complexity of coincident systems is the danger of increased imperfection-sensitivity which may erode factors of safety if not accounted for. This problem may become critical during the course of erection.

No general rules are proposed here to deal with the problem of simultaneous buckling for specific structural systems, we have merely pointed out the possible complications which may arise. If there is some doubt as to the post-buckling behaviour of a given system it may well be wise to avoid a coincident buckling situation if this is possible. Again, the philosophy of simultaneous mode design in structural optimization could well be re-examined in the light of the present results and rules established to decide when the philosophy is sound and when it is not.

References

1 Thompson, J. M. T. 'Basic principles in the general theory of elastic stability', *J. Mech. Phys. Solids,* vol.11 (1963), pp.13–20.

2 Koiter, W. T. 'On the stability of elastic equilibrium', Dissertation, Delft, Holland (1945).

3 von Karman, T. and Tsien, H. S. 'The buckling of thin cylindrical shells under axial compression', *J. Aero Sciences,* vol.8 (1941), no.8, p.302.

4 Thompson, J. M. T. and Supple, W. J. 'Erosion of optimum designs by compound branching phenomena', *J. Mech. Phys. Solids,* vol.21 (1973), pp.135–144.

5 Chilver, A. H. and Johns, K. C. 'Multiple path generation at coincident branching points', *Int. J. Mech. Sci.,* vol.13 (1971), pp.899–910.

6 Chilver, A. H. 'Coupled modes of elastic buckling', *J. Mech. Phys. Solids,* vol.15 (1967), pp.15–28.

7 Supple, W. J. 'Coupled branching configurations in the elastic buckling of symmetric structural systems', *Int. J. Mech. Sci.,* vol.9 (1967), pp.97–112.

8 Supple, W. J. 'Initial post-buckling behaviour of a class of elastic structural systems', *Int. J. Non-Linear Mechanics,* vol.4 (1969), pp.23–26.

9 Roorda, J. 'The buckling behaviour of imperfect structural systems', *J. Mech. Phys. Solids,* vol.13 (1965), p.267.

10 Supple, W. J. 'On the change in buckle pattern in elastic structures', *Int. J. Mech. Sci.,* vol.10 (1968), pp.737–745.

11 Supple, W. J. 'Coupled buckling modes of structures', Ph.D. Thesis, London University (1966).

4 Frame instability

J. W. Butterworth*

1 Introduction

Frames, both plane and three-dimensional, provide one of the commonest forms of construction in civil engineering and as a result their structural behaviour has been widely studied. In particular the problem of elastic stability of plane frames, and more recently of space frames, has been the subject of much attention. Early work on frames concentrated on purely local instabilities such as minor axis stanchion buckling, torsional-flexural buckling, flange and web buckling and lateral buckling of beam elements. The increasing slenderness of frames resulting from improved design methods, in particular plastic design, led naturally to more importance being attached to elastic stability problems.

The first serious attempt to study the overall elastic stability of frames was probably made by Lundquist [1] using a moment distribution method, modified to take account of the effect of axial forces on the stiffness and carryover factors of each member. The relationship between the axial force in a beam element and its flexural stiffness has been defined by means of stability functions, the best known tabulation of which is probably that prepared by Livesley and Chandler [2]. Using these stability functions Livesley [3] formulated a systematic analysis method for the elastic critical load of a plane frame in which an approximation to the stiffness matrix of the structure was formed for an increasing range of applied loading. The critical load was taken as that load above which the stiffness matrix was no longer positive definite.

A number of subsequent formulations were based on the same principle and varied only in their numerical analysis details. A recent text [4] describes the approach in some detail. Such an approach presumes the structure to remain undeformed under increasing load with only the axial forces in the members changing; consequently the only type of buckling behaviour that can be detected is bifurcation from the undeformed state, and although a limited class of plane frames (for example, plane rigidly jointed triangulated frames and symmetrically loaded portal type frames) do conform closely with this assumed buckling behaviour, many other possibilities may be overlooked.

More recent approaches to frame buckling [5, 6] have incorporated prebuckling deformation and other nonlinear effects and frequently use an iterative nonlinear analysis technique to follow the nonlinear equilibrium path up to the point of initial buckling. The details of such a technique have been explained clearly by Dickie (see Chapter 7) and require no further comment at this point.

* Department of Civil Engineering, University of Surrey

Clearly, whatever approach is adopted to investigate frame stability, it must be borne in mind that the behaviour of every conceivable frame must be in accordance with the general theory described in previous chapters (provided, of course, that only conservative elastic systems are being considered). This being the case it appears reasonable to assume that frames are capable of exhibiting the whole spectrum of elastic stability phenomena, including limit point, bifurcations and coupled mode behaviour, and methods of analysis should accordingly take all these possibilities into account.

2 Experimentally observed behaviour

The correlation between the predictions of the general theory and the physically observed behaviour of plane frames has been convincingly demonstrated by a series of experiments on model frames carried out by Roorda [7, 8]. The members of the models were made of high tensile steel strip with a cross-section of 25 mm by 1.0 to 1.5 mm, connected together rigidly by slotted dural joints. The proportions of the models and the strength of the steel ensured elastic behaviour even in the post-buckling range and the flat strip cross-section prevented out of plane buckling.

The first model tested consisted of two members connected together at right angles and supported on hinged joints (Fig.1). The load Λ was applied at the joint by means of a semi-rigid loading system through a knife edge which could be moved laterally to introduce small load eccentricities e. The rotation of the joint, θ, was used as a generalized coordinate. As the load was increased it was found that the joint rotation was either clockwise or anticlockwise according to whether e was greater or less than a particular value e_0. Fig.2(a) shows two sets of typical equilibrium paths for $e > e_0$ and $e < e_0$. With $e < e_0$, Λ increased to a maximum value Λ_{max} at which point snap buckling occurred in an anticlockwise mode. For $e > e_0$, a completely stable equilibrium path resulted, with steadily increasing clockwise deformation. By loading the structure until snap buckling occurred, then increasing the load slightly and at the same time forcing the structure into the opposite (clockwise) mode of deformation the complementary equilibrium path was obtained. The load was then slowly reduced until the structure eventually snapped back into the counter clockwise mode. Fig.2(b) shows the maximum snap-buckling loads reached for various values of eccentricity. The value e_0 could be regarded as the imperfection in loading necessary to 'cancel out' the initial geometric imperfections in the structure (assuming that the perfect system was regarded as one which became unstable at a point of bifurcation). It is clear that the paths in Fig.2(a) correspond to those of an imperfect asymmetric bifurcation point and the imperfection sensitivity curve of Fig.2(b) (of which only the lower half was obtained) agrees closely with the form predicted by the general theory. A second test using an inclined load was carried out on the frame shown in Fig.1. A similar set of results were obtained, with $e_0 < 0$ in this case. Figs 3(a) and 3(b) summarize the results and this time it may be observed that the complete set of snap buckling loads for a range of imperfection was obtained. Values of $(\Lambda_{max}/\Lambda_{cr}) > 1.0$ correspond to the points on the complementary equilibrium paths at which snap-back to an anti-

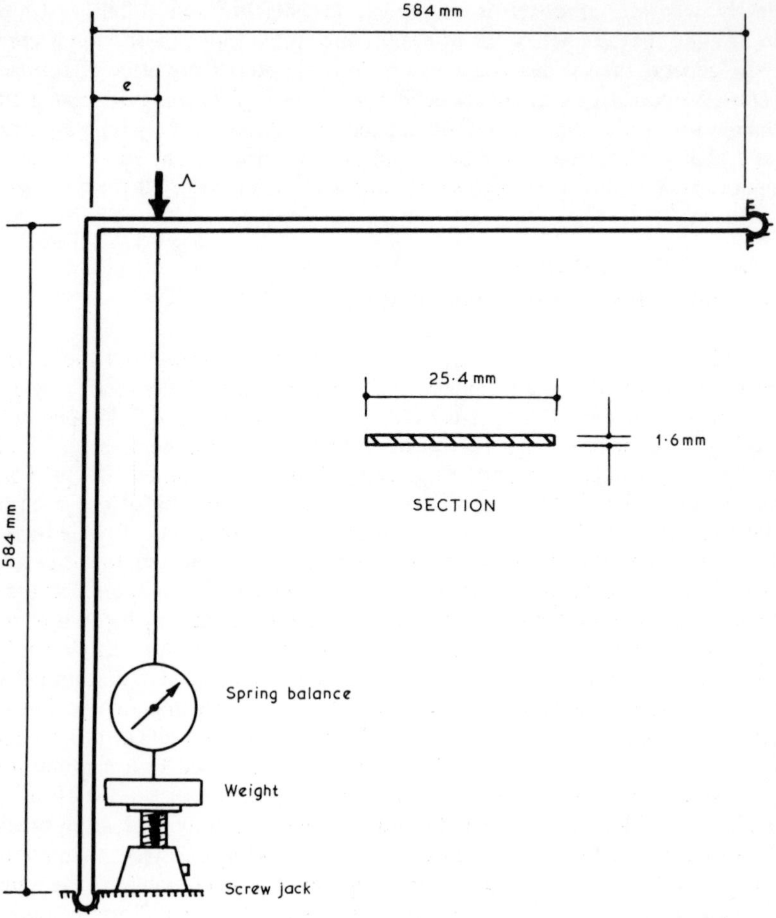

Fig. 1

clockwise mode occurred. Again it is easy to visualize an asymmetric point of bifurcation 'sandwiched' between the imperfect equilibrium paths of Fig.3(a).

Figs 4(b) and 4(c) summarize the results of tests on the shallow two-hinged frame shown in Fig.4(a). The span was 610 mm and the two identical members were of 25.4 mm by 1.0 mm cross-section. Load was applied at varying eccentricity in a similar way to the previous example. The type of behaviour in this case is obviously different from the previous two cases in certain respects. These differences are associated with the symmetry of the system which results in the buckling behaviour also being symmetric. Whether or not e is greater than e_0 the structure becomes unstable at a limit point, the mode being clockwise if $e > e_0$

or anticlockwise if $e < e_o$. Complementary paths were obtained by forcing the frame into a mode opposite to its natural inclination and increasing the load. These paths were all completely unstable, but of interest in that they helped to 'sandwich' the experimentally unobtainable perfect path corresponding to $e = e_o$. This perfect path can be seen from Fig.4(b) to be a fundamental path defined by $\theta = 0$, reaching an unstable-symmetric point of bifurcation at a load of Λ_{cr}. The plot showing the variation of critical snap-buckling loads with eccentricity (Fig.4(c))

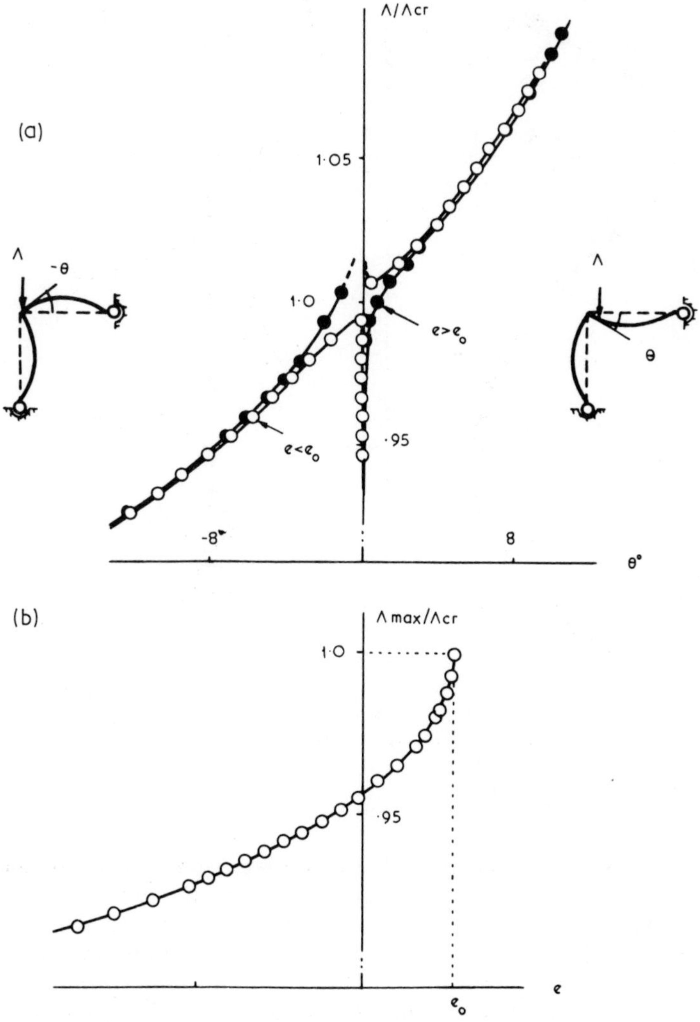

Fig. 2

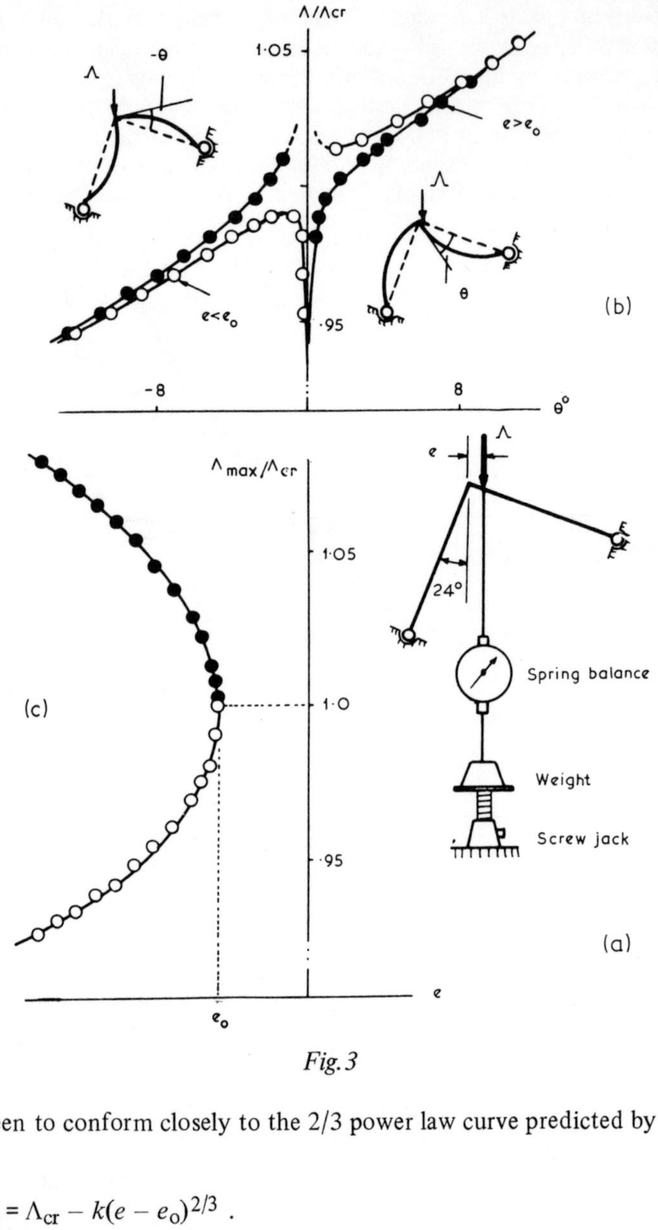

Fig. 3

can be seen to conform closely to the 2/3 power law curve predicted by the general theory

$$\Lambda_{max} = \Lambda_{cr} - k(e - e_o)^{2/3} \ . \tag{1}$$

A simple model that illustrates the *stable*-symmetric type of bifurcation is the simple strut illustrated in Fig.5(a). Consisting of a pin-ended steel strip with similar proportions to the members of the frame shown in Fig.1, loading was again by means of a semi-rigid loading system with provision for applying the load at a controlled eccentricity at one end. The two branches of the equilibrium path shown in Fig.5(b) are for an eccentricity $e < e_0$ and it can be seen that the branch of the curve in the right-hand quadrant (the complementary path) represents equilibrium states reached by forcing the strut into the mode of buckling opposite to that resulting from following the 'natural' loading path which in this case is the completely stable path shown in the left hand quadrant (θ negative). The complementary path reaches a limit point at a value of $\Lambda = \Lambda_{min}$ and ceases to exist at loads less than Λ_{min} . Λ_{min}

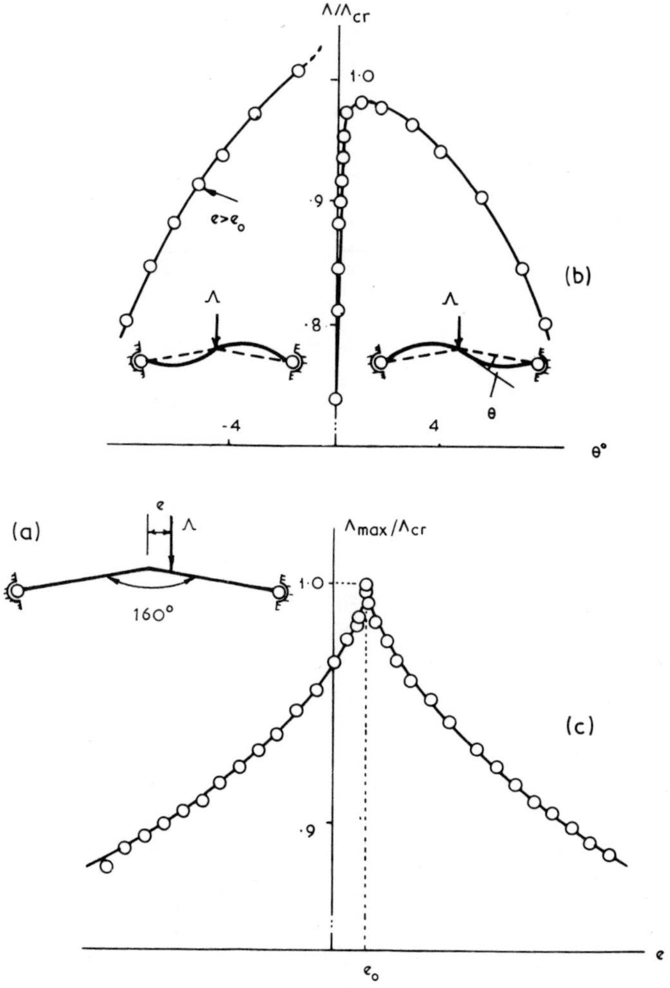

Fig.4

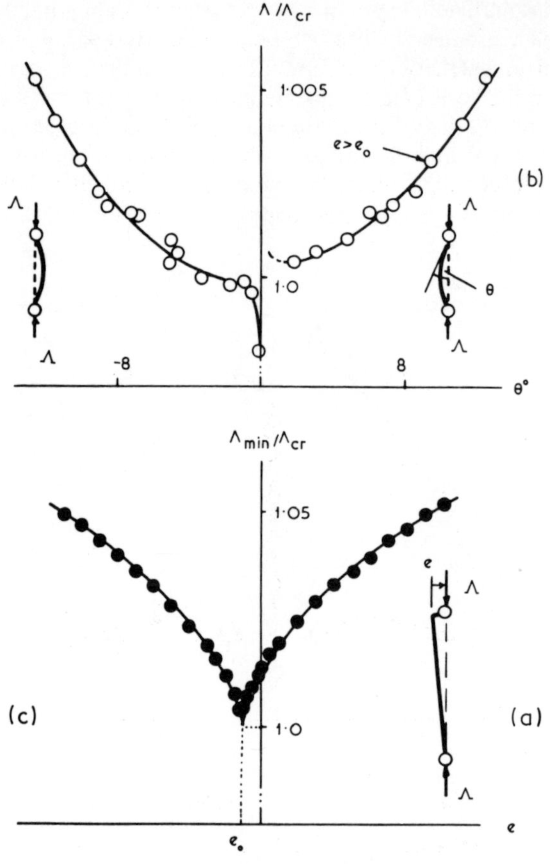

Fig. 5

varies with eccentricity e in the manner shown in Fig.5(c), in which the form of the curve is again seen to agree remarkably with the theoretical 2/3 power law of the general theory.

A test on the more complicated, rigidly jointed frame shown in Fig.6, loaded by symmetrically placed point loads $\Lambda/2$ with controlled eccentricities e, gave the results shown in Figs 7(a) and 7(b). The rotation θ of the top right-hand joint was chosen as a generalized coordinate and when plotted against load Λ for eccentricities greater than e_0 and less than e_0 gave rise to the two pairs of natural and complementary equilibrium paths shown. The sensitivity of the frame to imperfections was severe as can be seen from the curve of Fig.7(b) showing the variation in the snap-buckling loads of the imperfect systems with changing eccentricity. The form

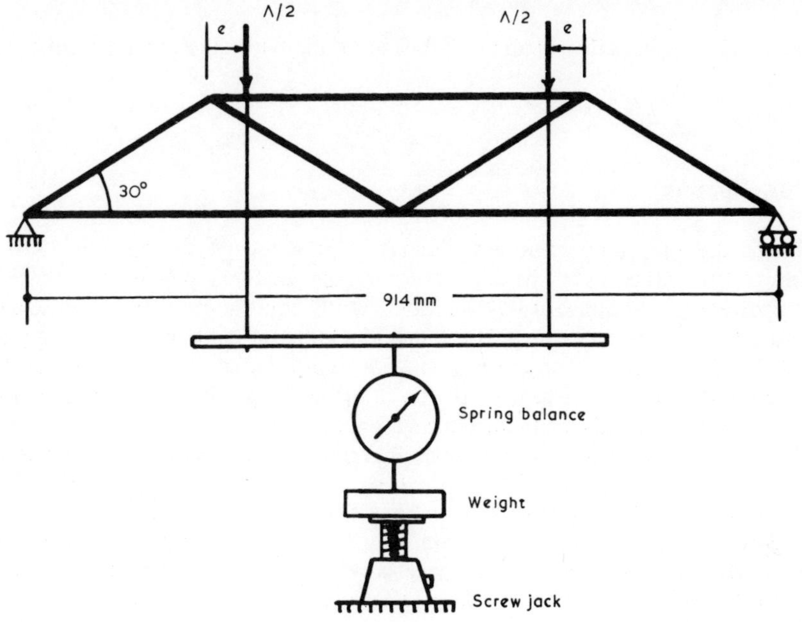
Λ/2
Λ/2
e
e
30°
914 mm
Spring balance
Weight
Screw jack

Fig. 6

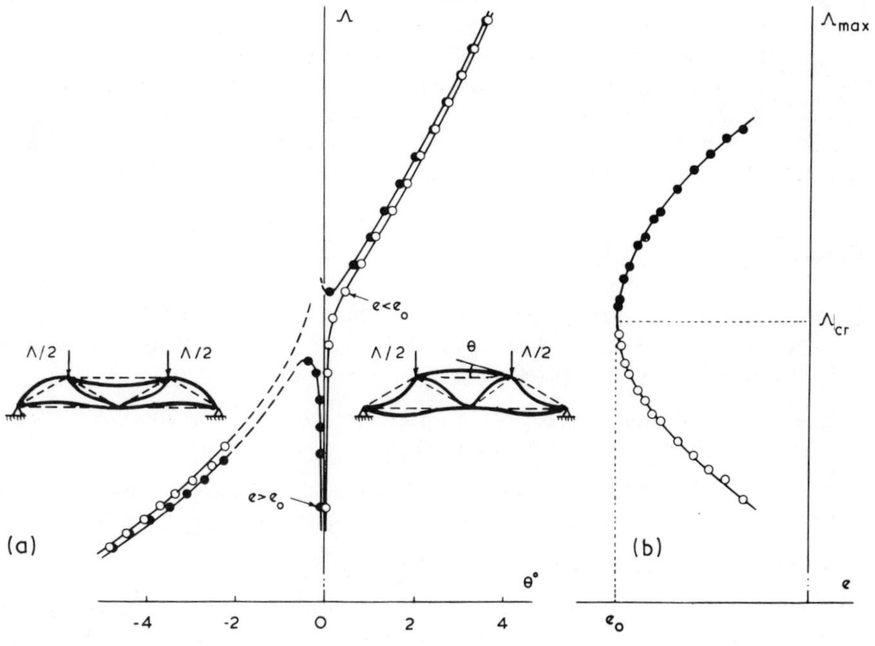
Λ
Λ/2
Λ/2
θ
e<e₀
e>e₀
(a)
θ°
-4
-2
0
2
4
Λmax
Λcr
(b)
e
e₀

Fig. 7

of the Λ_{max}-e curve is in agreement with that predicted by the general theory

$$\Lambda_{max} = \Lambda_{cr} \pm k(e - e_o)^{1/2} . \tag{2}$$

3 Conclusions

Although the preceding examples illustrate just a few of the possible forms of buckling of frames it is obvious that a considerable variety of pre- and post-buckling behaviour exists. Nonlinearity of the fundamental equilibrium path is noticeable in all cases, leading to either limit point snap-buckling or stable post-buckling paths. It follows that a rigorous theoretical analysis should be capable of following these nonlinear paths up to and beyond the point of initial buckling. For systems which bifurcate asymmetrically or in unstable symmetric form the effect of imperfections is highly significant, especially for small imperfections in the vicinity of the perfect system.

It should be noted, however, that such systems although sensitive to small imperfections may in some cases be comparatively insensitive to larger imperfections; in other words the imperfection sensitivity curve may have a steep or even infinite slope near zero imperfection but level off as imperfections increase, leading to a useful 'safe' buckling load.

In practice, the proportions of a structure and the limitations in the elastic properties of the construction material will mean that plasticity effects may become apparent as deformations increase, either before or after theoretical elastic buckling occurs. This possibility only serves to underline the necessity for accurate nonlinear *elastic* analysis, for how else can the true material stresses be found so that it may be known when yield stress has been reached? Having established that yield occurs in a particular structure it will obviously be necessary to undertake a more elaborate elastic-plastic type of analysis beyond the point of first yield [9].

References

1 Lundquist, E. E. 'Principles of moment distribution applied to stability of structural members', 5th International Conference of Applied Mathematics, Cambridge, Mass. (1952) p.145.

2 Livesley, R. K. and Chandler, D. B. *Stability Functions for Structural Frameworks,* Manchester University Press (1956).

3 Livesley, R. K. 'The application of an electronic digital computer to some problems of structural analysis', *The Structural Engineer* (Jan. 1956), pp.1–12.

4 Majid, K. I. *Non-Linear Structures,* Butterworths (1972).

5 Turner, M. J., Dill, E. H., Martin, H. C., Melosh, R. J. 'Large deflections of structures subjected to heating and external loads', *Journal of the Aerospace Sciences,* vol.27 (1960), pp.97–106.

6 Jennings, A. 'Frame analysis including change of geometry', *Journal of the Structural Division, ASCE,* vol.94 (Sep. 1968), no.ST9, Proc. Paper 6115, pp.627–644.

7 Roorda, J. 'Stability of structures with small imperfections', *Journal of the Engineering Mechanics Division, ASCE,* vol.91 (Feb. 1965), EM1, pp.87–106.

8 Roorda, J. 'The buckling behaviour of imperfect structural systems', *J. Mech. Phys. Solids,* vol.13 (1965), pp.267–280.

9 Sewell, M. J. Ch.5 in Leipholz, H. H. E. (Ed.), *Plastic Buckling,* SM Study no.6, University of Waterloo (1972).

5 Post-buckling behaviour of thin plates

W. J. Supple*

1 Introduction

The post-buckling behaviour of thin plates is an important topic in structural mechanics since plates are possibly unique in their extensive use as load-carrying structural components up to and into the post-buckling range. The axial stiffness of a plate after buckling is approximately one half of the pre-buckled value, the exact value being dependent upon the conditions at the plate boundaries and upon the number of axial halfwaves of the buckled form. Tests on plates in axial compression have shown that the waveforms adopted at the onset of buckling may undergo abrupt changes further into the post-buckling regimes. Associated with these abrupt changes in waveform will be corresponding changes in axial stiffness. When loading is continued above the critical value there is, besides the possible abrupt changes in longitudinal waveform, a gradual transformation of the transverse waveform characterized by a flattening of the central regions of the plate. This phenomenon with its associated re-distribution of axial mid-surface stresses has given rise to 'effective width' concepts and has been studied in some detail by Koiter [1].

The ability of plates loaded in edge compression to sustain loads well above the classical critical load would imply that such plates have post-buckling characteristics of the stable-symmetric type demonstrated in structural bifurcation theory. We shall show here that this is indeed so and furthermore that abrupt changes in waveform after initial buckling may be adequately explained in terms of nonlinear coupling of buckling modes at simultaneous or near-simultaneous critical loads. Apart from any possible fatigue effects due to repeated abrupt modification of waveform arising from cyclic loading in the post-buckling regime this nonlinear coupling apparently does not impair the post-buckling strength of the plate. As such, plates under in-plane loading have always been considered as being well-behaved structural components. The same cannot automatically be said of structures composed of assemblages of plates since such structures may adopt the buckling characteristics associated with thin shells and display unstable post-buckling behaviour. Furthermore, there may arise interaction or coupling between local buckling of individual plate elements and some overall buckling mode of the structure with resulting adverse effects on the load-carrying capacity of the structure.

Here we shall be interested in the elastic post-critical behaviour of individual plates under uni-axial in-plane compression since this loading condition may arise

* Department of Civil Engineering, University of Surrey.

in plate components of more complex thin-walled structures. We are examining in detail, therefore, the local buckling behaviour of these structures. It is felt that information on this component behaviour is a useful pre-requisite to any global investigation.

2 The plate post-buckling equations

Let us consider the problem of a thin rectangular plate loaded in its plane as shown in Fig.1 and which is simply-supported on all four edges in the conventional sense and which has further specified restrictions on the boundary displacements. The mode of support is such that there are no out-of-plane deflections at the boundaries, the loaded edges remain straight and the longitudinal edges are not allowed to wave in the plane of the plate. The latter condition applies to a single panel of a multi-panelled infinitely wide plate loaded in axial compression, the junctions of the panels being knife-edge supports. It is further assumed that there is no restraint against lateral expansion of the plate in its plane. Adopting the plate dimensions and co-ordinate axes as shown in Fig.1 these boundary conditions may be written as

$$\begin{aligned}
&\text{(i)} \quad (w)_{x=a/2;-a/2} = 0,\\
&\text{(ii)} \quad (w)_{y=b/2;-b/2} = 0,\\
&\text{(iii)} \quad (w_{,xx} + \nu w_{,yy})_{x=a/2;-a/2} = 0,\\
&\text{(iv)} \quad (w_{,yy} + \nu w_{,xx})_{y=b/2;-b/2} = 0,\\
&\text{(v)} \quad (u)_{x=a/2;-a/2} = \text{const.}\\
&\text{(vi)} \quad (v)_{y=b/2;-b/2} = \text{const.}
\end{aligned} \tag{1}$$

where a comma followed by subscripts represents partial differentiation in turn with respect to each subscripted variable. No restrictions are placed upon the other boundary displacements which will be discussed later together with results for other boundary conditions.

The behaviour of plates loaded beyond the critical load is governed by the von-Kármán large deflection equations which relate the out-of-plane buckling deflections w to the in-plane forces per unit width N_x, N_y, N_{xy}. If we consider also the presence of initial geometric imperfections w^0 these equations can be written in terms of w, w^0 and a stress function ϕ in the form

$$\begin{aligned}
&\nabla^4\phi = E\{(w_{,xy})^2 - w_{,xx}w_{,yy} - [(w^0_{,xy})^2 - w^0_{,xx}w^0_{,yy}]\},\\
&\nabla^4(w - w^0) = \frac{t}{D}\{\phi_{,yy}w_{,xx} - 2\phi_{,xy}w_{,xy} + \phi_{,xx}w_{,yy},
\end{aligned} \tag{2}$$

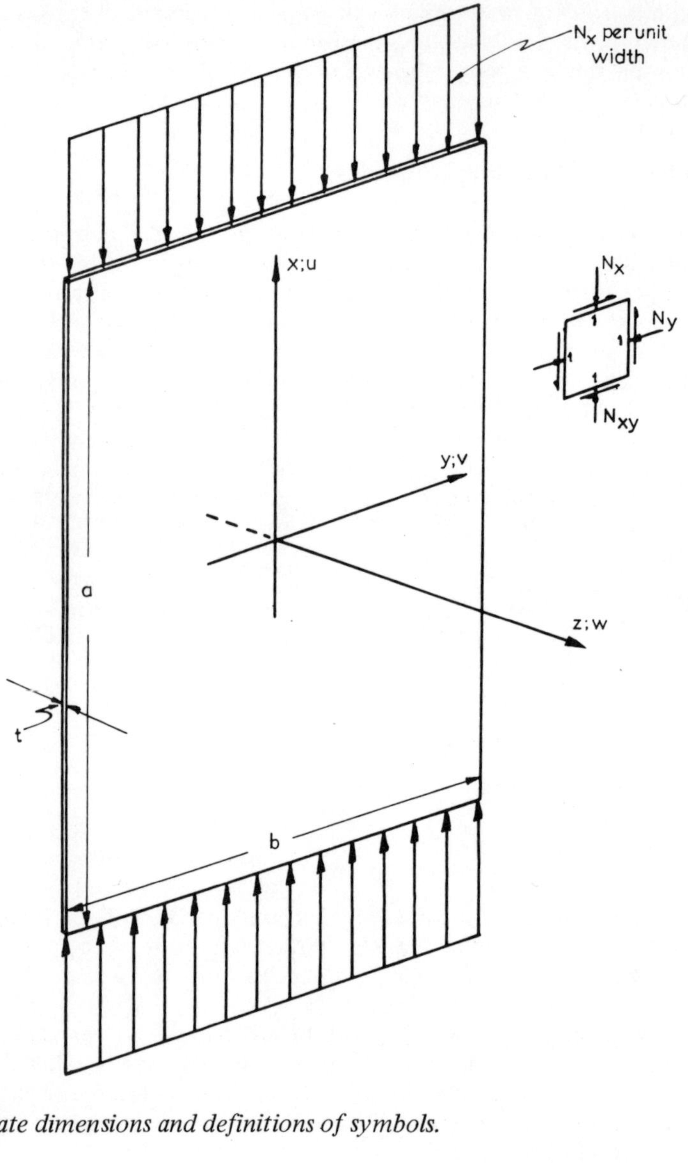

Fig. 1 Plate dimensions and definitions of symbols.

in which w represents the total out-of-plane deflection from the flat configuration and where ϕ is defined as

$$\phi_{,xx} = \frac{N_y}{t}; \quad \phi_{,yy} = \frac{N_x}{t}; \quad \phi_{,xy} = -\frac{N_{xy}}{t}. \tag{3}$$

The first of these equations is a statement of the conditions of compatibility of the in-plane strains and the second equation represents the condition of equilibrium of forces in the out-of-plane or z direction. These statements are clarified in Fig.2 which is a schematic layout of the information used in deriving these equations for the perfect plate.

For a given loading and set of boundary conditions solution of these simultaneous equations to establish the buckled form of a plate is extremely complicated. For the present discussion we shall resort to an approximate solution of these equations using a Galerkin technique by assuming a buckled form in accord with experimental observations. The post-buckling equations which are obtained could also be derived by substituting the assumed form for w and its associated quantities into the expression for the total potential energy of the plate and then expressing the equilibrium conditions as stationary values of this energy in the usual way. However, it is found that this entails more algebraic effort than in the method employed here.

Experiments show that when a thin rectangular plate is supported at its edges and loaded in its plane above the critical load it adopts a buckled configuration which has the form of one halfwave across the plate width and a number of halfwaves along its length which is dependent upon the length to width ratio γ (aspect ratio) of the plate. The detailed form of this buckling mode depends upon the type of support afforded at the plate boundaries.

With the origin of coordinates at the plate centre as shown in Fig.1 we may

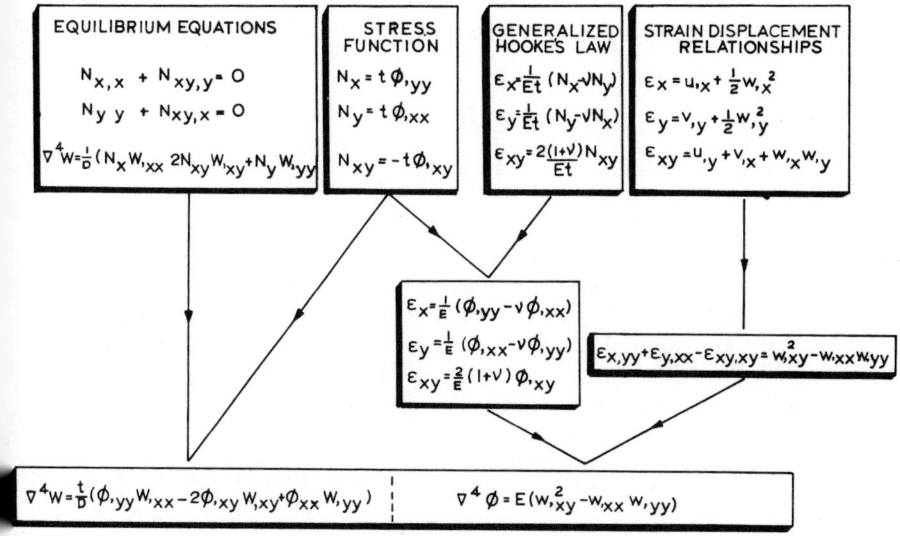

Fig. 2 The von-Kármán large deflection equations.

satisfy the simple support boundary conditions (1) by a choice of w in the form

$$w = A\cos\frac{\pi y}{b}\cos\frac{r\pi x}{a} \tag{4}$$

where we have effectively reduced the problem to that of a single degree of freedom represented by buckling in r halfwaves longitudinally (in the x direction). This choice of w would be sufficient to establish the stable-symmetric nature of the post-buckling of plates and also to predict the post-buckling response in the absence of other effects for buckling in r halfwaves. Careful experiments have shown [2] that accurate values of out-of-plane deformation can be obtained with such a simple formulation for values of load up to at least two and a half times the critical load, for very thin plates. However, we find that the critical loads for buckling in m and n longitudinal halfwaves become coincident when the plate aspect ratio $\gamma = \sqrt{mn}$ and since the general theory of elastic stability predicts the possibility of nonlinear coupling of buckling modes when such coincidence occurs we investigate such effects in the sequel by choosing

$$w = \cos\frac{\pi y}{b}\left(A\cos\frac{n\pi x}{a} + B\sin\frac{m\pi x}{a}\right) \tag{5}$$

and further we select initial geometric imperfections in the form

$$w^0 = \cos\frac{\pi y}{b}\left(A_0\cos\frac{n\pi x}{a} + B_0\sin\frac{m\pi x}{a}\right). \tag{6}$$

With this choice for w we may establish the mode interaction for short plates ($\gamma < 4$) and tentatively extend the results to explain changes of waveform in the post-buckling range for long plates. Long plates have many coincident or near-coincident critical loads and it may be argued that a two-mode analysis may fail to detect some types of behaviour. However, changes of mode form even for long plates normally occur between two distinct mode shapes.

There is an apparent restriction on the evenness or oddness of m and n in (5) and (6) in order that boundary condition (1(i)) be satisfied: however, since the post-buckling equations which result are interchangeable in (n, A) and (m, B) this restriction (which is merely a consequence of the selected position for the origin of the coordinate axes) becomes irrelevant. These expressions are substituted into the first of equations (2) (the compatibility equation) and the resulting equation solved for ϕ to give

$$\phi = -\frac{E}{32}(A^2 - A_0^2)\left(\frac{a^2}{n^2b^2}\cos\frac{2n\pi x}{a} + \frac{n^2b^2}{a^2}\cos\frac{2\pi y}{b}\right)$$

$$-\frac{Ea^2}{4b^2}(AB - A_0B_0)\left(\frac{1}{(m+n)^2}\sin\frac{(m+n)\pi x}{a} + \frac{1}{(m-n)^2}\sin\frac{(m-n)\pi x}{a}\right)$$

$$-\frac{Ea^2b^2}{4}\cos\frac{2\pi y}{b}(AB - A_0B_0)\left(\alpha\sin\frac{(m+n)\pi x}{a} + \beta\sin\frac{(m-n)\pi x}{a}\right)$$

$$+\frac{E}{32}(B^2 - B_0^2)\left(\frac{a^2}{m^2b^2}\cos\frac{2m\pi x}{a} + \frac{m^2b^2}{a^2}\cos\frac{2\pi y}{b}\right) - \frac{\lambda}{2bt}y^2, \qquad (7)$$

in which λ represents the total applied end load

$$\int_{-b/2}^{+b/2} N_x \mathrm{d}y$$

and where α and β are defined as

$$\alpha = \frac{(m-n)^2}{\{(m+n)^2b^2 + 4a^2\}^2}, \quad \beta = \frac{(m+n)^2}{\{(m-n)^2b^2 + 4a^2\}^2}. \qquad (8)$$

The midsurface stresses may be derived from this expression for ϕ and thence by use of the strain-displacement relationships and the generalized Hooke's Law expressions may be obtained for u and v in terms of the midsurface stresses and w. When this is done it is found that u and v satisfy the boundary conditions (1(v)) and (1(vi)) provided $(m + n)$ is an odd integer, which will always be so for the cases considered herein. The variation of u along the longitudinal edges and the variation of v across the loaded edges may also be calculated from these expressions; it is assumed, however, that such movements may occur and are not restrained.

Next the expressions for w and ϕ are substituted into the second of equations (2) (the out-of-plane equilibrium equation) which will not be satisfied exactly but will have a residual R (say) owing to the approximate nature of w. The integral over the surface of the plate

$$\int_{-a/2}^{+a/2}\int_{-b/2}^{+b/2} Rw \,\mathrm{d}x\,\mathrm{d}y \qquad (9)$$

has the dimensions of work and we define this as the excess energy of the plate. The Galerkin approximation requires this virtual excess energy to vanish which leads to the two simultaneous equations

$$\int_{-a/2}^{+a/2}\int_{-b/2}^{+b/2} R \cos\frac{n\pi x}{a} \cos\frac{\pi y}{b}\, \mathrm{d}x\, \mathrm{d}y = 0,$$

$$\int_{-a/2}^{+a/2}\int_{-b/2}^{+b/2} R \sin\frac{m\pi x}{a} \cos\frac{\pi y}{b}\, \mathrm{d}x\, \mathrm{d}y = 0, \tag{10}$$

which are nonlinear algebraic equations in A, B, A_0 and B_0. Performing the substitutions and integrations, equations (10) appear as

$$\frac{3(1-\nu^2)a^2b^2}{16n^2}\left[\left(\frac{n^4}{a^4}+\frac{1}{b^4}\right)(\bar{A}^2-\bar{A}_0^2)+\frac{m^2n^2}{a^4}(\bar{B}^2-\bar{B}_0^2)\right]\bar{A}$$

$$+\,12(1-\nu^2)\frac{a^2b^2}{n^2}\left(K-\frac{m^2n^2}{64a^4}\right)(\bar{A}\bar{B}-\bar{A}_0\bar{B}_0)\bar{B}+\frac{a^2b^2}{4n^2}\left(\frac{n^2}{a^2}+\frac{1}{b^2}\right)^2(\bar{A}-\bar{A}_0)$$

$$-\frac{\lambda b}{4\pi^2 D}\bar{A}=0,$$

$$\frac{3(1-\nu^2)a^2b^2}{16m^2}\left[\left(\frac{m^4}{a^4}+\frac{1}{b^4}\right)(\bar{B}^2-\bar{B}_0^2)+\frac{m^2n^2}{a^4}(\bar{A}^2-\bar{A}_0^2)\right]\bar{B}+12(1-\nu^2)\frac{a^2b^2}{m^2}$$

$$\times\left(K-\frac{m^2n^2}{64a^4}\right)(\bar{A}\bar{B}-\bar{A}_0\bar{B}_0)\bar{A}+\frac{a^2b^2}{4m^2}\left(\frac{m^2}{a^2}+\frac{1}{b^2}\right)^2(\bar{B}-\bar{B}_0)-\frac{\lambda b}{4\pi^2 D}\bar{B}=0, \tag{11}$$

where

$$K=\frac{1}{16b^4}+\frac{1}{64}[(m-n)^2\alpha+(m+n)^2\beta]+\frac{m^2n^2}{64a^4}, \tag{12}$$

and where

$$\bar{A}=\left(\frac{A}{t}\right),\quad \bar{B}=\left(\frac{B}{t}\right),\quad \bar{A}_0=\left(\frac{A_0}{t}\right),\quad \bar{B}_0=\left(\frac{B_0}{t}\right). \tag{13}$$

Following the general theory for two-degree-of-freedom systems we define an uncoupled buckling mode as a solution of equations (11) which involves only one of the deformation parameters $\bar{A}$ or $\bar{B}$ and we make the following observations:

(i) when $\bar{A}_0 = \bar{B}_0 = 0$, there are three solutions of equations (11); two are uncoupled, the third coupled;

(ii) when $\bar{A}_0 \neq 0, \bar{B}_0 = 0$ or $\bar{A}_0 = 0, \bar{B}_0 \neq 0$, there are two solutions of equations (11); one is uncoupled, the other coupled;

(iii) when $\bar{A}_0 \neq 0$, $\bar{B}_0 \neq 0$, there is one solution of equations (11) and this is coupled.

As was implied earlier the uncoupled buckling modes for the ideal plate are stable-symmetric in character when plotted in the λ versus $\bar{A}$, $\bar{B}$ planes. Supple [3, 4] has made a detailed study in general terms of the possible forms of the coupled buckling modes for structural systems with two degrees of freedom which have symmetric uncoupled modes. If we denote the uncoupled mode with the lower critical load as the primary mode and that with the higher critical load as the secondary mode then the results of [3] may be summarized using the notation of the present analysis as follows. The form of the $\bar{A}$ vs $\bar{B}$ relationship of the coupled mode for perfect symmetrical structural systems, when real, may be either

(i) an ellipse, implying a transition path from the primary to the secondary mode;

(ii) a hyperbola branching from the primary mode, implying a further instability phenomenon after initial buckling;

(iii) a hyperbola branching from the secondary mode, implying that if the initial post-buckling is stable it will remain so.

Bearing these findings in mind we now examine the solutions of equations (11) to determine the form of mode coupling present for the particular boundary conditions under consideration. Later we shall see that the boundary conditions influence greatly the nature of the coupling.

3 The ideal plate

The equations governing the behaviour of a perfectly flat plate are obtained by putting $\bar{A}_0 = \bar{B}_0 = 0$ in equations (11) which then reduce to the forms

$$\bar{A}\left\{12(1-\nu^2)\frac{a^2b^2}{n^2}\left[\frac{1}{64}\left(\frac{n^4}{a^4}+\frac{1}{b^4}\right)\bar{A}^2+K\bar{B}^2\right]+\frac{a^2b^2}{4n^2}\left(\frac{n^2}{a^2}+\frac{1}{b^2}\right)^2-\frac{\lambda b}{4\pi^2 D}\right\}=0,$$

$$\bar{B}\left\{12(1-\nu^2)\frac{a^2b^2}{m^2}\left[\frac{1}{64}\left(\frac{m^4}{a^4}+\frac{1}{b^4}\right)\bar{B}^2+K\bar{A}^2\right]+\frac{a^2b^2}{4m^2}\left(\frac{m^2}{a^2}+\frac{1}{b^2}\right)^2-\frac{\lambda b}{4\pi^2 D}\right\}=0, \quad (14)$$

from which the uncoupled buckling modes are derived as the solutions (i) $\bar{A} \neq 0$, $\bar{B} = 0$ and (ii) $\bar{A} = 0, \bar{B} \neq 0$ or explicitly as

$$\frac{\lambda b}{4\pi^2 D} = \frac{a^2 b^2}{4n^2}\left(\frac{n^2}{a^2} + \frac{1}{b^2}\right)^2 + \frac{12(1-\nu^2)}{64n^2}\, a^2 b^2 \left(\frac{n^4}{a^4} + \frac{1}{b^4}\right)\bar{A}^2,$$

$$\frac{\lambda b}{4\pi^2 D} = \frac{a^2 b^2}{4m^2}\left(\frac{m^2}{a^2} + \frac{1}{b^2}\right)^2 + \frac{12(1-\nu^2)}{64m^2}\, a^2 b^2 \left(\frac{m^4}{a^4} + \frac{1}{b^2}\right)\bar{B}^2. \qquad (15)$$

The coupled solution by definition is represented by the solution for which $\bar{A} \neq 0$, $\bar{B} \neq 0$ and the $\bar{A}$-$\bar{B}$ relationship for this situation may be obtained by elimination of λ between equations (14) and appears in the form

$$\left\{\frac{1}{64n^2}\left(\frac{n^4}{a^4} + \frac{1}{b^4}\right) - \frac{K}{m^2}\right\}\bar{A}^2 + \left\{\frac{K}{n^2} - \frac{1}{64m^2}\left(\frac{m^4}{a^4} + \frac{1}{b^4}\right)\right\}\bar{B}^2 =$$

$$= \frac{1}{48(1-\nu^2)}\left\{\frac{1}{m^2}\left(\frac{m^2}{a^2} + \frac{1}{b^2}\right)^2 - \frac{1}{n^2}\left(\frac{n^2}{a^2} + \frac{1}{b^2}\right)^2\right\}. \qquad (16)$$

We can see by inspection that, as anticipated, the individual buckling modes represented by equations (15) display stable-symmetric post-buckling equilibrium paths. Examination of equation (16) will indicate where any secondary bifurcations may appear on these paths. The plate aspect ratio $\gamma = a/b$ and so if for argument $m > n$ then we have the conditions that if $\gamma^2 < mn$ the uncoupled mode with n halfwaves is the primary mode and if $\gamma^2 > mn$ the uncoupled mode with m halfwaves is the primary mode. We also note that the right-hand side of equation (16) is negative when $\gamma^2 > mn$ and positive when $\gamma^2 < mn$ and that the coefficient of $\bar{A}^2$ in this equation is always negative and the coefficient of $\bar{B}^2$ always positive. As a result we conclude that *the form of equation (16) is a hyperbola and the coupled mode always branches from the secondary mode.* For loads greater than that at which the coupled mode branches from the secondary mode the general theory proves that the latter is stable.

The post-buckling equilibrium paths for an ideal plate with $\gamma = 2, m = 3, n = 2$, $\nu = 1/3$ are shown in Fig.3. It may be noted that in equations (15) $(4\pi^2 D)/b$ is the classical buckling load of a simply-supported plate whose aspect ratio is an integer and accordingly the first terms on the right-hand sides represent the buckling stress coefficient (normally referred to in the texts as k_c) divided by four. We may calculate the critical loads at which the coupled mode bifurcates from the secondary mode and thus define what we might term the coupled buckling stress coefficient ${}^c k_c$ related to the buckling stress coefficient as follows:

$$^{c}k_c = k_c + \left\{ \frac{\left(\dfrac{n^2b^2}{a^2} + \dfrac{a^2}{n^2b^2}\right)\left[\dfrac{1}{m^2}\left(\dfrac{m^2}{a^2} + \dfrac{1}{b^2}\right)^2 + \dfrac{1}{n^2}\left(\dfrac{n^2}{a^2} + \dfrac{1}{b^2}\right)^2\right]}{\dfrac{1}{n^2}\left(\dfrac{n^4}{a^4} + \dfrac{1}{b^4}\right) - \dfrac{64K}{m^2}} \right\} \tag{17}$$

This inter-relationship is illustrated in Fig.4 for varying aspect ratios: the values of $^{c}k_c$ are shown by the dash-dot curves. It can be seen that as γ increases the critical loads for buckling in the uncoupled and coupled modes rapidly converge.

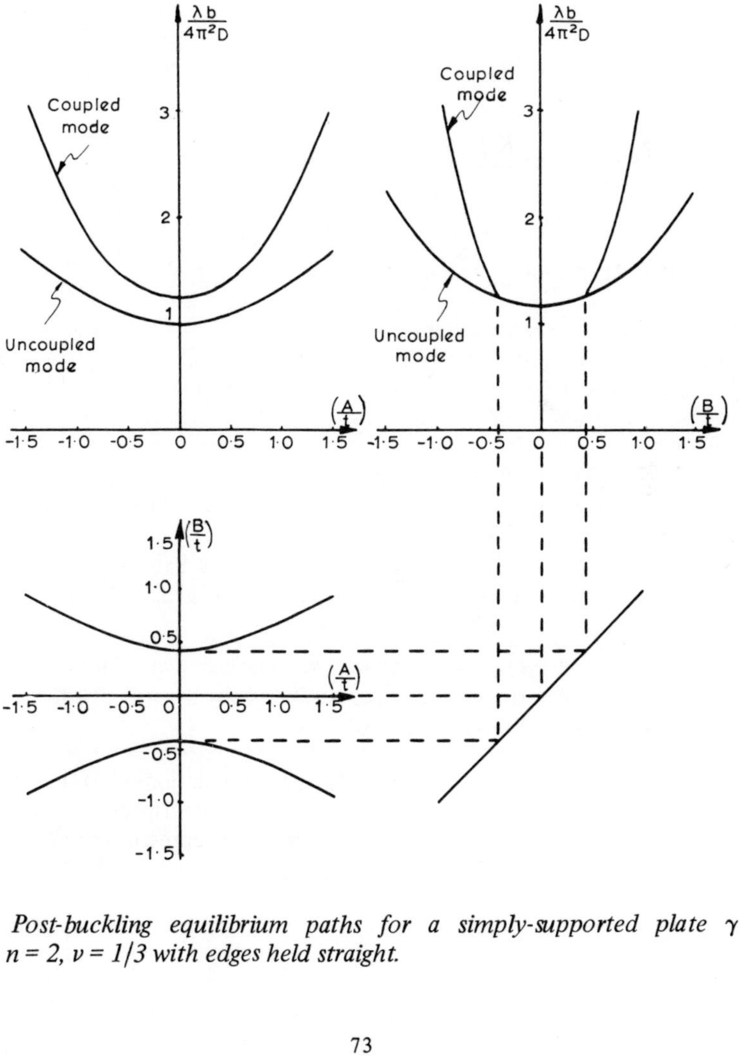

Fig.3 Post-buckling equilibrium paths for a simply-supported plate γ = 2, m = 3, n = 2, ν = 1/3 with edges held straight.

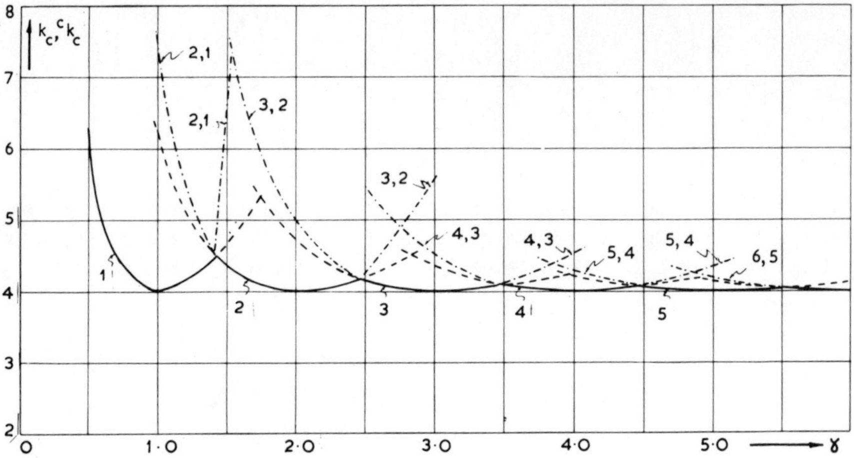

Fig. 4 Critical loads for coupled and uncoupled modes of buckling.

4 Initial imperfections

All structures will have inherent imperfections in form and the general theory of elastic stability demonstrates that such imperfections will have a profound effect upon the stability behaviour of any structure. A stability analysis investigating post-buckling behaviour would be incomplete without inclusion of the effects of initial imperfections. Geometric imperfections in the shape of the natural buckling modes or with large components of these quantities have the most marked effect upon the post-buckling structural response. This explains the choice of the imperfections given by equation (6).

If, for convenience, we consider that buckling in n halfwaves longitudinally represents the primary buckling mode for a plate then we may divide the buckling of imperfect plates into three classes. These are

(i) imperfection in primary mode, $\bar{A}_0 \neq 0, \bar{B}_0 = 0$;
(ii) imperfection in primary mode, $\bar{A}_0 = 0, \bar{B}_0 \neq 0$;
(iii) imperfections in both modes, $\bar{A}_0 \neq 0, \bar{B}_0 \neq 0$.

From the general theory of [4] we know that when there is only an initial imperfection in the form of the primary mode then the plate will deform into this mode and no interaction from the secondary mode occurs. For example, if we prescribe $m = 3, n = 2, \nu = 1/3$ and $\bar{B}_0 = 0$ then the first of equations (11) becomes

$$0.333\bar{A}(\bar{A}^2 - \bar{A}_0^2) + 1.395\bar{A}\bar{B}^2 + (\bar{A} - \bar{A}_0) - \frac{\lambda b}{4\pi^2 D}\bar{A} = 0 \tag{18}$$

and the $\bar{A}$-$\bar{B}$ relationship of the coupled mode is

$$2.50\bar{A}_0 + 0.434\bar{A} + 0.416\bar{A}\bar{A}_0^2 - 2.375\bar{A}\bar{B}^2 + 0.719\bar{A}^3 = 0. \tag{19}$$

Fig.5 shows the post-buckling equilibrium paths obtained from equations (18) and (19) when $\bar{A}_0 = 1.0$; the equilibrium paths for the perfect plate are shown as broken lines. Besides the natural loading path, which the plate follows when loaded, all the complementary paths are shown for the sake of completeness.

A value $\bar{A}_0 = 1.0$ is a very large imperfection and was chosen here to produce good separation between the equilibrium paths illustrated in Fig.5 bearing in mind that the qualitative behaviour demonstrated would be the same for all values of $\bar{A}_0$ in the range 0 to 1.0.

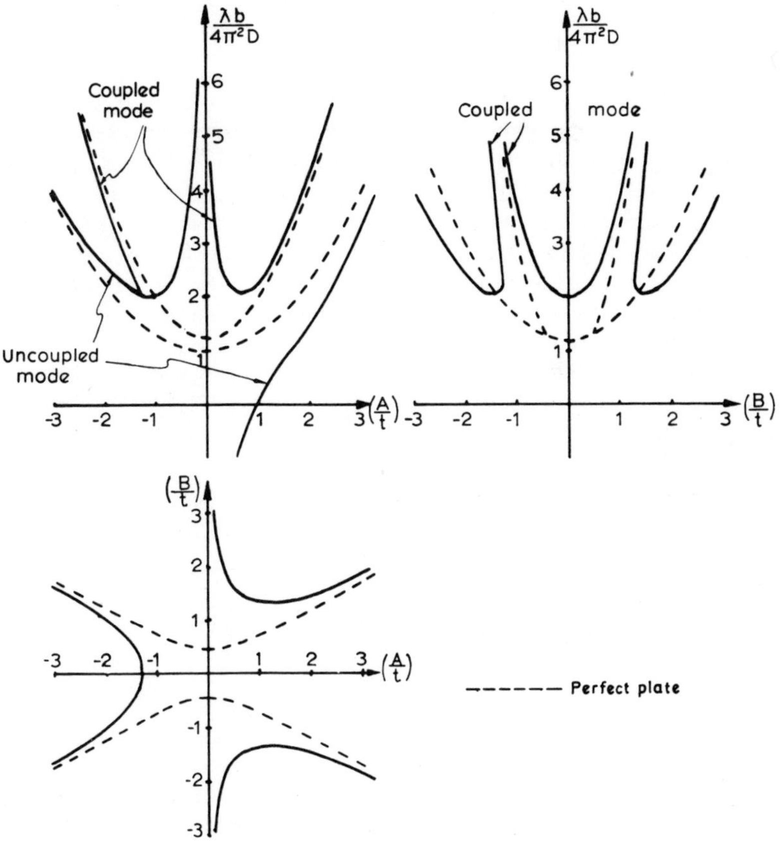

Fig.5 Post-buckling equilibrium paths for imperfect plate $\gamma = 2$, $m = 3$, $n = 2$, $\nu = 1/3$ with initial imperfection $\bar{A}_0 = 1.0$.

When only an initial imperfection in the form of the secondary mode is present, i.e. $\bar{A}_0 = 0$, then with $m = 3, n = 2, \nu = \frac{1}{3}$ the first of equations (11) reduces to the form

$$0.333\bar{A}^2 + 1.395\bar{B}^2 - 0.375\bar{B}_0^2 + 1.00 - \frac{\lambda b}{4\pi^2 D} = 0, \tag{20}$$

and the $\bar{A}$-$\bar{B}$ relationship of the coupled mode is now

$$2.94\bar{B}_0 - 0.434\bar{B} - 0.72\bar{A}^2\bar{B} + 0.185\bar{B}\bar{B}_0^2 + 2.37\bar{B}^3 = 0. \tag{21}$$

When $\bar{A} = 0$ equation (21) may have one or three real roots for $\bar{B}$ depending on the magnitude of the imperfection $\bar{B}_0$. That value of $\bar{B}_0$ that makes two of the roots real and equal represents a critical imperfection [4] and from equation (21) this value for the assumed magnitudes of γ, ν, m, n is

$$(\bar{B}_0)_{crit} = 0.025.$$

If $\bar{B}_0 > (\bar{B}_0)_{crit}$ the buckled form of the plate develops as three half-sinewaves and there is no interaction from the primary two-halfwave mode. If $\bar{B}_0 < (\bar{B}_0)_{crit}$ the plate initially develops in the three-half-sinewave mode but this mode then becomes unstable at a point of bifurcation whereupon the plate buckles into a coupled mode which on further loading approaches asymptotically to the primary two-half-sinewave mode. These points are illustrated in Figs 6 and 7; in Fig.6 the initial imperfection $\bar{B}_0 = 0.020$ and in Fig.7 $\bar{B}_0 = 1.00$.

Finally, when both initial imperfections $\bar{A}_0$ and $\bar{B}_0$ are non-zero, the buckling equations for the arbitrarily assigned values $m = 3, n = 2, \nu = \frac{1}{3}$ appear as

$$A - A_0 + 0.333(\bar{A}^2 - \bar{A}_0)\bar{A} + 0.375(\bar{B}^2 - \bar{B}_0^2)\bar{A} + 1.02(\bar{A}\bar{B} - \bar{A}_0\bar{B}_0)\bar{B} - \frac{\lambda b}{4\pi^2 D} = 0,$$

$$0.294\bar{A}\bar{B}_0 - 0.25\bar{B}\bar{A}_0 - 0.434\bar{A}\bar{B} - 0.071\bar{A}^3\bar{B} + 0.237\bar{A}\bar{B}^3 + 0.0185\bar{A}\bar{B}\bar{B}_0^2 -$$

$$- 0.0416\bar{A}\bar{B}\bar{A}_0^2 + 0.112\bar{A}^2\bar{A}_0\bar{B}_0 - 0.255\bar{B}^2\bar{B}_0\bar{A}_0 = 0. \tag{22}$$

The forms of the solution to these equations are shown in Fig.8 for two combinations of initial imperfections $\bar{A}_0$ and $\bar{B}_0$. It is clear from the results shown that for a given value of A_0 there is a critical value of $\bar{B}_0$ for which the stability of the natural loading path is lost at a point of bifurcation; when $\bar{A}_0 = 0$ this corresponds to the critical value of $\bar{B}_0$ already discussed. By ignoring second order terms of $\bar{A}_0$ and $\bar{B}_0$ in equations (11) we may derive the approximate locus of critical imperfections (i.e. combinations of $\bar{A}_0$ and $\bar{B}_0$ which produce a branching solution of the natural loading path) in the form

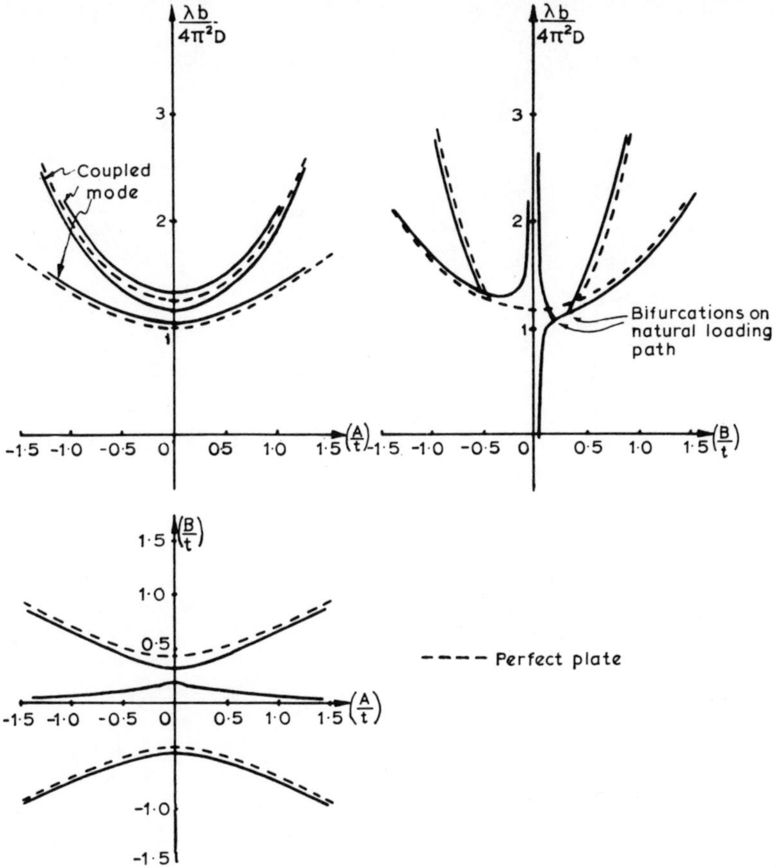

Fig. 6 Post-buckling equilibrium paths for imperfect plate $\gamma = 2$, $m = 3$, $n = 2$, $\nu = 1/3$ with initial imperfection $\overline{B}_0 = 0.020$.

$$K_1\overline{A}_0^{2/3} + K_2\overline{B}_0^{2/3} = K_3, \tag{23}$$

where

$$K_1 = \left\{12(1-\nu^2)\left[\frac{1}{64n^2b^4} - \frac{K}{m^2} + \frac{n^2}{64a^4}\right]\right\}^{1/3}\left\{\frac{1}{8n^2}\left(\frac{n^2}{a^2} + \frac{1}{b^2}\right)^2\right\}^{2/3},$$

$$K_2 = \left\{12(1-\nu^2)\left[\frac{K}{n^2} - \frac{1}{64m^2b^4} - \frac{m^2}{64a^4}\right]\right\}^{1/3}\left\{\frac{1}{8m^2}\left(\frac{m^2}{a^2} + \frac{1}{b^2}\right)^2\right\}^{2/3},$$

$$K_3 = \frac{(m^2-n^2)}{12m^2n^2}\left(\frac{m^2n^2}{a^4} - \frac{1}{b^4}\right). \tag{24}$$

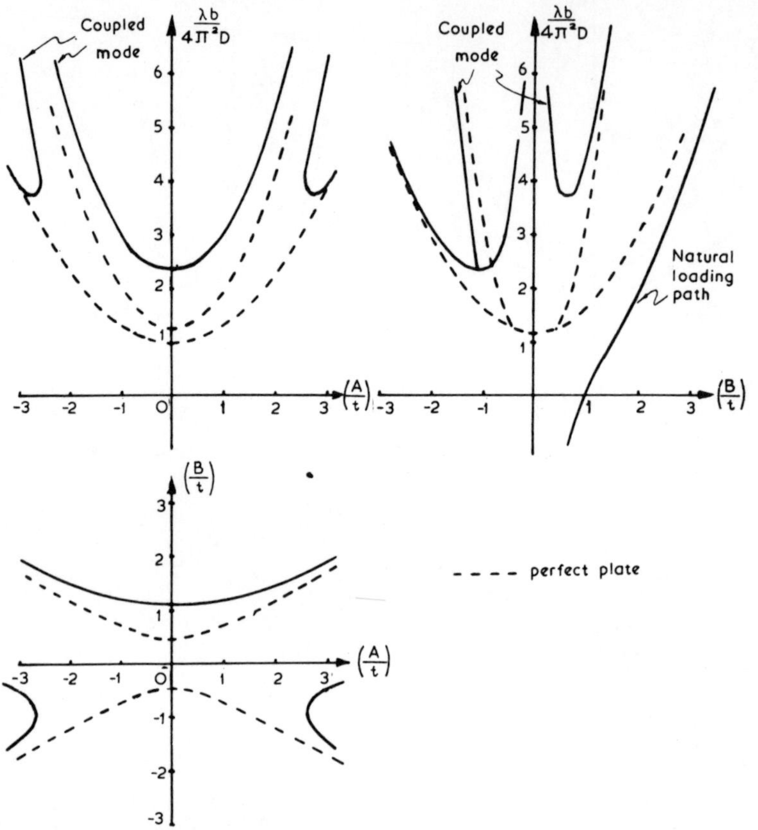

Fig. 7 Post-buckling equilibrium paths for imperfect plate $\gamma = 2$, $m = 3$, $n = 2$, $\nu = 1/3$ with initial imperfection $\bar{B}_0 = 1.0$.

This critical locus is shown plotted in Fig.8 for the appropriate values of γ, m, n, ν and divides the set of positive initial imperfections $(\bar{A}_0, \bar{B}_0)$ into two sub-sets. If the initial imperfection belongs to the sub-set containing the $\bar{A}_0$ axis then at loads of approximately twice the classical buckling load the buckling mode is predominantly two half-sine waves; if the initial imperfection belongs to the other sub-set then at these loads the buckling mode is predominantly three half-sinewaves. It might be emphasized again that in the absence of initial imperfections buckling would be in a purely two-half-sinewave mode.

5 Alternative boundary conditions

Let us now turn our attention to other support restraints along the unloaded edges of the compressed rectangular plate. It has been shown by Hlavacek [5] that for a simply-supported plate which is allowed to expand laterally and whose unloaded edges are free to wave in the plane of the plate the coupled buckling mode is of the hyperbola type branching from the upper uncoupled buckling mode. Again, Supple [2] has shown the same type of coupling behaviour to exist for the simply-supported plate which is free to expand laterally and whose unloaded edges are clamped.

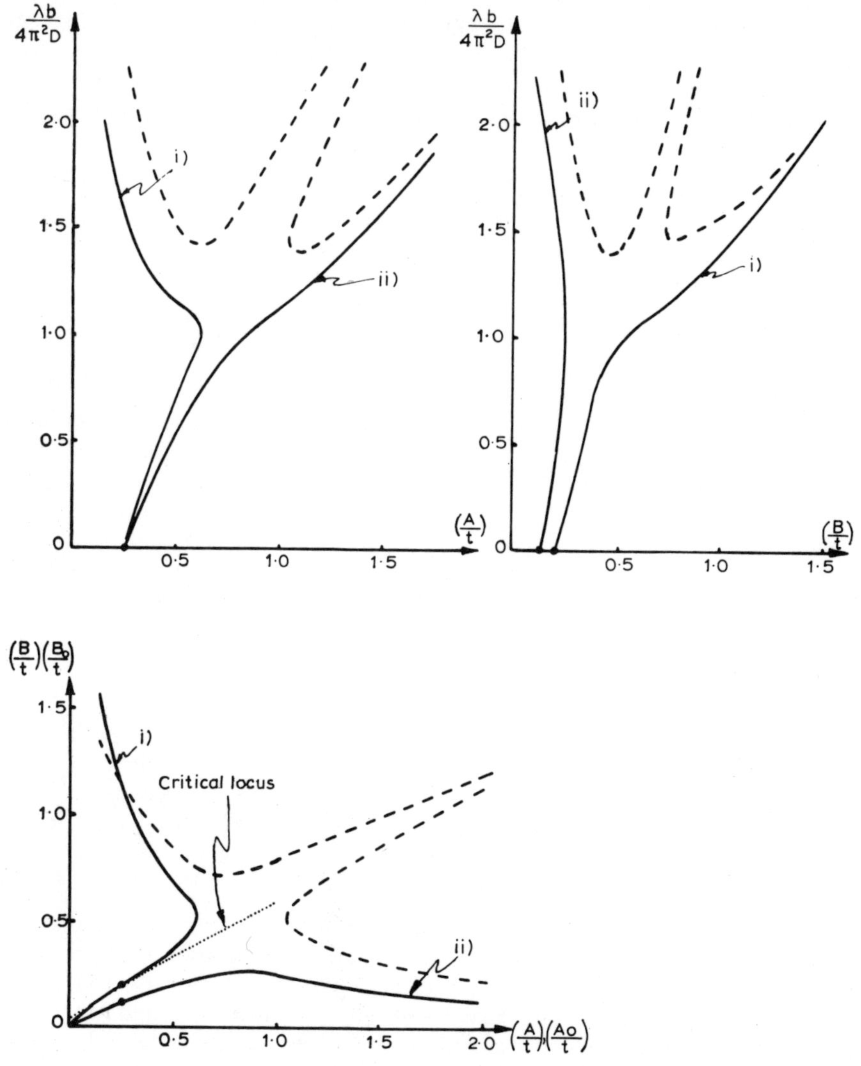

Fig. 8 Natural loading paths for a plate with $\gamma = 2$, $m = 3$, $n = 2$, $\nu = 1/3$ and initial imperfections (i) $\bar{A}_0 = 0.25$, $\bar{B}_0 = 0.20$; (ii) $\bar{A}_0 = 0.25$, $\bar{B}_0 = 0.125$.

Therefore, for these two types of boundary conditions the post-buckling behaviour of plates is qualitatively the same as that described in the preceding sections of the present analysis.

Major changes in waveform after initial buckling for these boundary conditions would be explained by the presence of initial imperfections.

If the plate is restrained from expanding laterally then the form of the coupled buckling mode is dependent upon the value of Poisson's ratio of the plate material. This observation follows from the results of Sharman and Humpherson [6] who analysed the buckling behaviour of a simply-supported plate whose unloaded edges are rigidly held apart. In the terminology of the present report the buckling equations corresponding to equations (11) obtained by these authors appear in the forms

$$\bar{B}\left\{12(1-\nu^2)\left[\left(\frac{1+3\gamma^4}{16}\right)\bar{B}^2+Q_1\bar{A}^2\right]+(1+\gamma^2)^2-\frac{\lambda a^2}{\pi^2 Db}(1+\gamma^2\nu)\right\}=0,$$

$$\bar{A}\left\{12(1-\nu^2)\left[\left(\frac{16+3\gamma^4}{16}\right)\bar{A}^2+Q_1\bar{B}^2\right]+(4+\gamma^2)^2-\frac{\lambda a^2}{\pi^2 Db}(4+\gamma^2\nu)\right\}=0, \tag{25}$$

where

$$Q_1=\tfrac{1}{8}(2+3\gamma^4)+\frac{81\gamma^4}{16(1+4\gamma^2)^2}+\frac{\gamma^4}{16(9+4\gamma^2)^2}, \tag{26}$$

and where buckling in two and one half-sinewaves only has been considered corresponding to $\bar{A}$ and $\bar{B}$ respectively. The $\bar{A}$-$\bar{B}$ relationship for the coupled mode follows from equations (25) in the form

$$\left\{Q_1(4+\gamma^2\nu)-\frac{(16+3\gamma^4)}{16}(1+\gamma^2\nu)\right\}\bar{A}^2$$

$$+\left\{\frac{(1+3\gamma^4)}{16}(4+\gamma^2\nu)-Q_1(1+\gamma^2\nu)\right\}\bar{B}^2=\frac{(4-\gamma^4+5\gamma^2\nu+2\gamma^4\nu)}{4(1-\nu^2)}, \tag{27}$$

which when real represents an ellipse or hyperbola as anticipated. The results for the forms of the coupled buckling mode derived from equation (27) are summarized in Fig.9 which is a plot of aspect ratio γ against Poisson's ratio ν. The unbroken curve represents simultaneous buckling, i.e. the values of γ and ν at which the critical loads for buckling in one and two half-sinewaves are equal, this condition being given by the equation

$$4-\gamma^4+5\gamma^2\nu+2\gamma^4\nu=0. \tag{28}$$

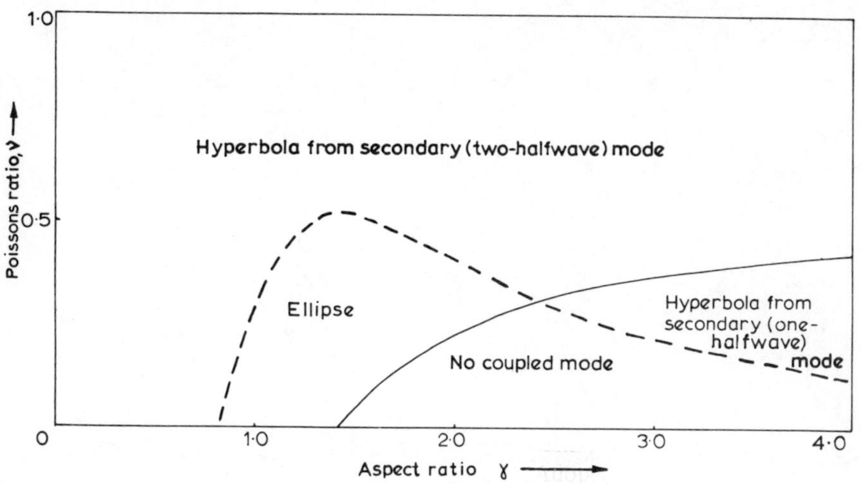

Fig.9 Forms of coupled buckling mode for a simply-supported plate with longitudinal edges rigidly held apart for m = 1, n = 2.

The lower right portion of the figure represents values of γ and ν for which buckling in two half-sinewaves is the primary mode. It may be noted that equation (28) is the same as the right-hand side of equation (27) indicating that there is a modification of the form of the coupled mode when the branching points of the uncoupled modes coalesce [3]. Any change in sign of the coefficients of $\bar{A}^2$ or $\bar{B}^2$ in equation (27) also represents a change in form of the coupling solution. The broken curve in Fig.9 represents the values of γ and ν at which the coefficient of $\bar{B}^2$ in equation (27) becomes equal to zero; the coefficient of $\bar{A}^2$ maintains the same sign for all positive values of γ and ν.

We see then that with these boundary conditions the closed transition or ellipse-type of coupled mode may be generated which is similar to the previously discussed form of coupling in that interchanges between the stable-symmetric uncoupled modes occur, but no overall instability is implied. The latter would occur only if coupling of the form of a hyperbola branching from the primary mode existed. This appears not to be so and explains the well-behaved load carrying properties of plates in the post-critical range.

As an example we may consider the plate with aspect ratio 2 and Poisson's ratio $^1/_3$. We see that the lowest critical load is associated with buckling in one half-sinewave which is thus the primary mode, and that the coupled mode is of the ellipse type. The load-deformation characteristics for this case are illustrated in Fig.10; it is seen that the plate initially buckles in one half-sinewave and then snaps to the two half-sinewave form at a load which is approximately 2.8 times the initial buckling value. With these boundary conditions, therefore, it is possible for the

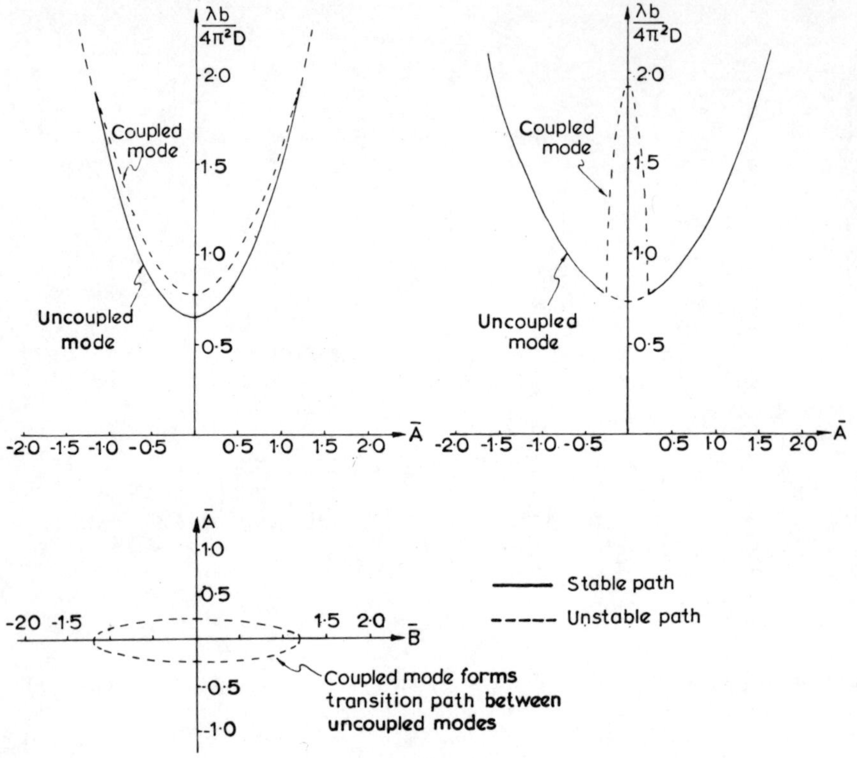

Fig. 10 Post-buckling equilibrium paths for a simply-supported plate $\gamma = 2$, $m = 1$, $n = 2$, $\nu = 1/3$ with longitudinal edges rigidly held apart.

ideal plate to undergo a change in buckle pattern provided the load mentioned is attained. Any initial imperfection with a two half-sinewave component would tend to reduce the load value at which the plate snaps from a predominantly one half-sinewave form to a predominantly two half-sinewave form.

6 Concluding remarks

This two-mode analysis has shown the complexities which may arise in the post-buckling behaviour of a simple, supposedly well-behaved structural component such as a rectangular plate in axial compression. The analysis confirmed the existence of types of mode coupling interaction predicted by the general theory of two-degrees-of-freedom doubly-symmetric systems. The results offer an explanation of the changes of waveform of compressed plates or plate elements in the post-critical range observed in experimental investigations. A multi-mode analysis would almost certainly arrive at the same conclusions. Whereas the results of the general theory are rigorously correct only in an asymptotic sense, that is, for vanishingly small deformations, the validity of the plate results here are dependent upon the range of validity of the von-Kármán large deflection equations and upon the accuracy of the approximate Galerkin technique used in their solution. It appears that the asymptotic general theory approach to post-buckling behaviour has missed no qualitative behaviour and has predicted all the phenomena witnessed here in the post-buckling of compressed plates.

References

1 Koiter, W. T. NLL Report S287, Amsterdam, December (1943).

2 Supple, W. J. 'Coupled buckling modes of structures', Ph.D. Thesis, London University (1966).

3 Supple, W. J. 'Coupled branching configurations in the elastic buckling of symmetric structural systems', *Int. J. Mech. Sci.*, vol.9 (1967), pp.97–112.

4 Supple, W. J. 'On the change of buckle pattern in elastic structures', *Int. J. Mech. Sci.*, vol.10 (1968), pp.737–745.

5 Hlavacek, I. 'Einfluss der Form der Anfangskrummung auf das Ausbeulen der Gedruckten Rechteckigen Platte', *Acta Tech. CSAV Prague,* vol.2 (1962), pp.174–206.

6 Sharman, P. W. and Humpherson, J. G. 'An experimental and theoretical investigation of simply-supported thin plates subjected to lateral load and uni-axial compression', *Aero. J. Roy. Aero. Soc.*, vol.72 (1968), pp.431–436.

6 Phenomenological aspects of some shell buckling problems

S. C. Tillman*

1 Introduction

Thin shell members are widely used in aerospace and marine structures where their high strength and low weight can be used to advantage. Their thinness however raises the problem of possible elastic buckling under compressive loads and a considerable amount of theoretical and experimental effort has been expended over the last ten years or so in an attempt to come to grips with this problem and to establish safe design criteria for practical shells. This effort is still continuing.

Many shell-loading combinations are found to exhibit small displacement linear (or near linear) pre-buckling behaviour shown by the line OA on a schematic load versus representative displacement plot in Fig.1. Initial theoretical studies treated the buckling of these shells as a linear eigenvalue problem; that is attention was focussed on finding point B on the line OA where the shell could just sustain a buckled form. This type of analysis is capable of yielding the buckling mode (or modes) of the shell but gives no information as to their subsequent stability. The load P_c corresponding to point B is referred to as the 'classical' buckling load. Other studies concentrated on the difficult nonlinear problem of determining the large displacement behaviour of the shell by using approximate energy methods together with plausible assumptions as to the shape of the post-buckling mode. The discovery from this work that many of the common shell-loading combinations possessed buckled equilibrium states at loads considerably lower than the classical (for example curve DEF in Fig.1) cast doubt on the suitability of using this load as a basis for design as it was clear that these shells would be highly sensitive to any imperfections in geometry or load application present in a prototype. Some recent very careful tests on near perfect shell-loading combinations, known to possess unstable post-buckling behaviour, have upheld these doubts and have shown that quite small imperfections can produce drastic reductions in the load-carrying capacity compared with that expected from the classical analysis. The curve OGH in Fig.1 represents the loading path of one of these imperfect shells. The shell would carry load until it reached the local maximum at G where (under force loading) it would snap to a highly buckled state at point H.

---

* Simon Engineering Laboratories, University of Manchester.

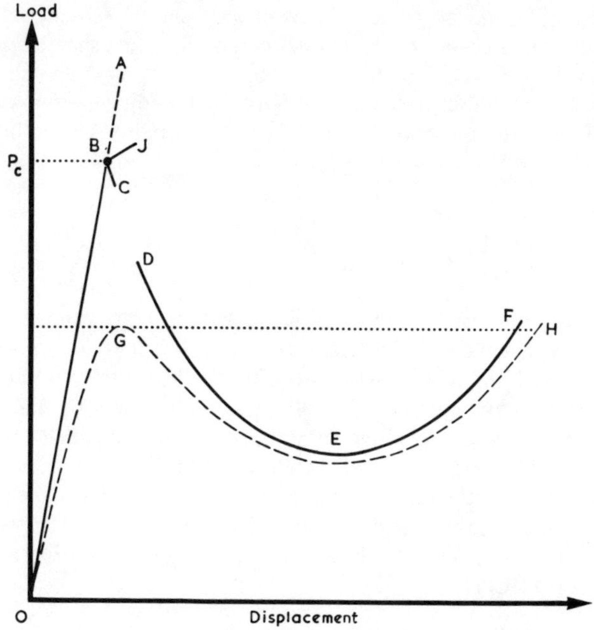

Fig. 1

From the above remarks it is obviously important, from the design point of view, to know if a particular shell-loading combination will be imperfection-sensitive and, if so, what reduction from the classical load can be expected for a given size of imperfection. This does not necessarily depend on a large deflection analysis as a knowledge of just the slope of the initial post-buckling path at the bifurcation point would be a useful indicator. If this is found to be rising (line BJ in Fig.1) imperfections should have little effect on the shell's behaviour. On the other hand if it is found to be falling (line BC in Fig.1) the shell will be imperfection-sensitive. Recent workers have followed this approach and have determined the initial post-buckling behaviour and imperfection-sensitivity of several common shell-loading combinations by exploiting the general theory of initial post-buckling behaviour first produced by Koiter [1]. Some results of this work for the particular cases of spherical and cylindrical shells (and some others that are closely related) are given in the rest of the text. These two shell types have been chosen because of their wide practical application. Some important aspects of the initial post-buckling theory are first briefly outlined to enable these results to be fully comprehended.

2 General theory of initial post-buckling behaviour

Consider a shell-loading combination having a linear elastic pre-buckling response and possessing a unique post-buckling mode. The perfect shell will exhibit one of the three types of initial post-buckling behaviour shown schematically in Fig.2. Here δ represents the amplitude of the buckling mode, t the shell's thickness and P the load. In the symmetrical cases (a) and (b) the post-buckling response is independent of the sign of the displacements. Many shell-loading combinations fall into this category. It is found from the theory [2] that the paths AB can be represented in an asymptotic sense (i.e. as $\delta/t \to 0$) by the series

$$\frac{P}{P_c} = 1 + a\frac{\delta}{t} + b\left[\frac{\delta}{t}\right]^2 + \ldots \tag{1}$$

where a represents the initial slope of these curves. For the symmetrical cases (a) and (b) this coefficient will be identically zero and the coefficient b will determine if the initial post-buckling path rises or falls after the bifurcation point. Clearly if b turns out to be negative the path is falling and consequently the shell will be imperfection-sensitive, and a positive value for b implies the shell retains its load-carrying capacity. If the shell is assumed to have an initial geometrical imperfection having the same shape as the buckling mode it will tend to follow the dotted curves shown in Fig.2. In this case Koiter has derived an asymptotic relationship connecting the maximum load P_s with the amplitude of the initial imperfection $\bar{\delta}$. For the symmetrical case (a) this is

$$\left[1 - \frac{P_s}{P_c}\right]^{3/2} = \frac{3\sqrt{3}}{2}\sqrt{-b}\left|\frac{\bar{\delta}}{t}\right|\frac{P_s}{P_c} \tag{2}$$

and is shown plotted for several values of b in Fig.3.

3 The axially loaded circular cylinder

The axially loaded circular cylinder has been notorious for many years for the great disparity that has existed between theoretical predictions of buckling strength and those observed experimentally. This difference is now known to be due to the

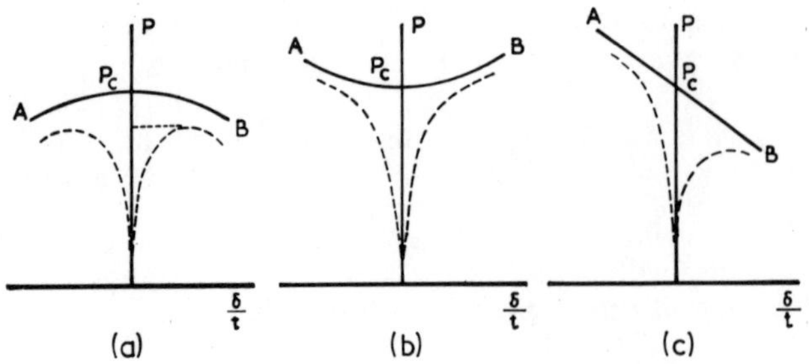

Fig. 2

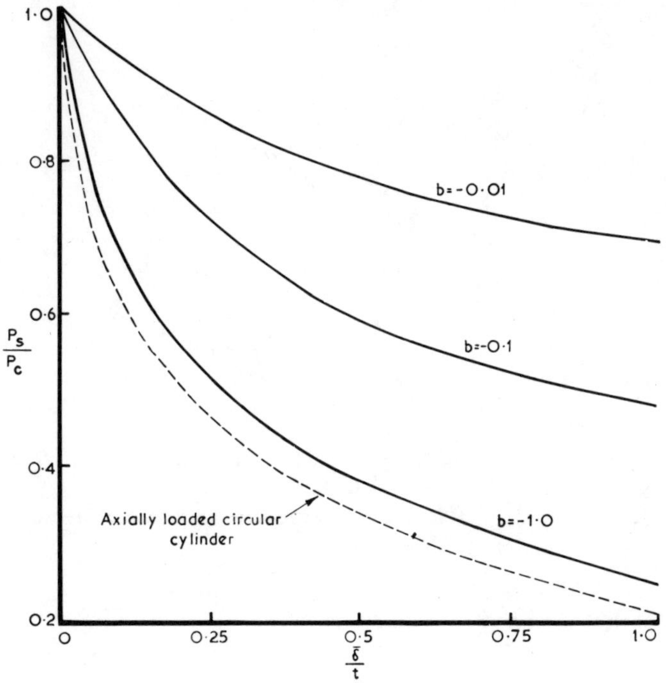

Fig. 3

extreme imperfection-sensitivity of this shell-loading combination. The classical buckling load of a perfect, infinitely long, cylinder is given in the notation of Fig.4(a) by

$$P_c = \frac{2\pi E t^2}{\sqrt{3(1 - \nu^2)}},$$

where E = Young's Modulus, ν = Poisson's ratio. This load is reduced somewhat when the length of the cylinder is taken into consideration. There is, however, no unique post-buckling mode corresponding to P_c. Several different modes are possible, all of which are unstable, and it is the coupling together of these modes, together with the effect of randomly distributed imperfections in practical shells, that leads to the large scatter in the test results. The extremely unstable post-buckling response of the perfect shell-loading combination is shown schematically in Fig.4(b) together with a probable path followed by the practical counterpart having small imperfections.

By assuming an axisymmetric sinusoidal buckling mode Koiter [3] has derived the curve shown in Fig.3 for the imperfection-sensitivity of the axially loaded circular cylinder. This curve is useful as it provides a rough calibration between

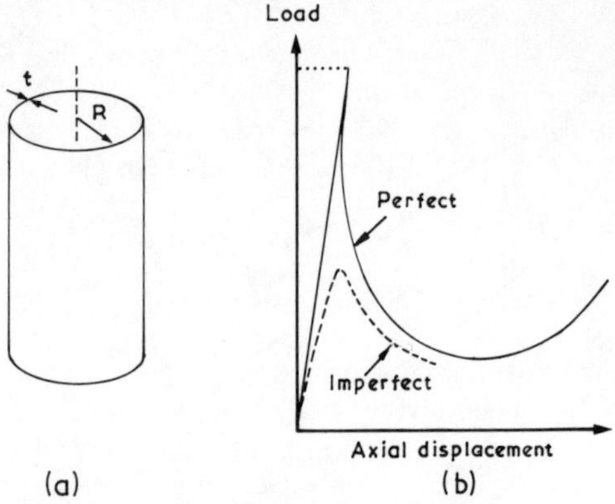

Fig.4

the degree of imperfection-sensitivity of structures displaying a unique post-buckling mode and the coefficient b as depicted in Fig.3. That is, a shell-loading combination having a value of b near to -1 would imply it is as imperfection-sensitive as the axially loaded circular cylinder.

4 Axially loaded oval cylindrical shells

These shells show unstable initial post-buckling behaviour [4] like the circular cross-sectioned cylinder but are of interest in that for certain values of cross-section eccentricities they appear to rapidly restabilize and are then able to sustain loads well above the classical before finally collapsing by snapping action [5]. The cross-section of an oval cylinder is shown in Fig.5(a). The geometry can be defined by two parameters R_s and q_0 such that

$$R_s = \frac{R_0}{1 - \xi \sin [R_s/R_0]} \quad \text{and} \quad q_0 = \sqrt[4]{12(1-\nu^2)} \, . \sqrt{\frac{R_0}{t}} \, ,$$

where

R_s = local circumferential radius of curvature

and

$$R_0 = \frac{\text{perimeter length of the oval}}{2\pi}$$

ξ is an 'eccentricity parameter' such that when $\xi = 0$ the cross-section is circular and when $\xi = 1$ the section corresponds to an oval having B/A = 0.485 with infinite radii of curvature at the ends of the minor axis. For a given value of q_0 the oval cylinder has a greater buckling load than the corresponding circular cylinder [5]. Fig.5(a) shows the value of the initial post-buckling coefficient b as a function of eccentricity (corresponding to a value of $q_0 = 18.2$) where it can be seen that the more eccentric the shells the less imperfection-sensitive they become. A large deflection analysis [5] has revealed however that when $0.5 \leqslant \xi \leqslant 0.8$ the post-buckling path is initially stable. The complete loading path shown in Fig.5(b) has been suggested [4] as the probable behaviour of the shells in this range.

5 Circular cylinders under pressure loading

The buckling and initial post-buckling behaviour of the pressure-loaded cylindrical shells shown in Fig.6 have been investigated in references [6] and [7]. In Fig.6(a) the shell is simply supported at the ends and in 6(b) it is periodically supported along its length by rigid ring frames. The critical loads for these two configurations are found to be similar [6]. An initial post-buckling analysis [7] has revealed the situation shown in Fig.6(c) where the coefficient b is plotted against the curvature parameter Z of the cylinder. When $1 < Z < 10$ the simply-supported cylinder is highly imperfection-sensitive but gets progressively less so as Z increases. This trend is confirmed by the test results [7]. For the ring-stiffened shell, however, maximum imperfection-sensitivity occurs when $Z \triangleq 10$ and for very low values of Z this shell configuration can even show a stable post-buckling response.

6 Complete spherical shell under external pressure

Many of the remarks made in connection with the axially loaded cylinder are equally applicable to the complete spherical shell under external pressure loading. The classical buckling pressure is given, in the notation of Fig.7, by

$$P_c = \frac{2E(t/R)^2}{\sqrt{3(1-\nu^2)}} \tag{3}$$

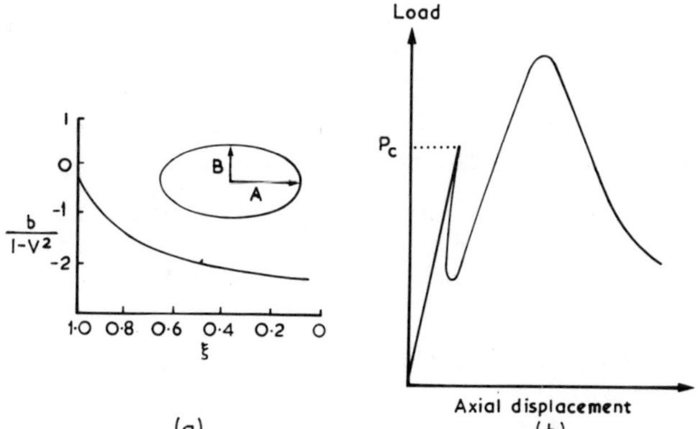

Fig. 5

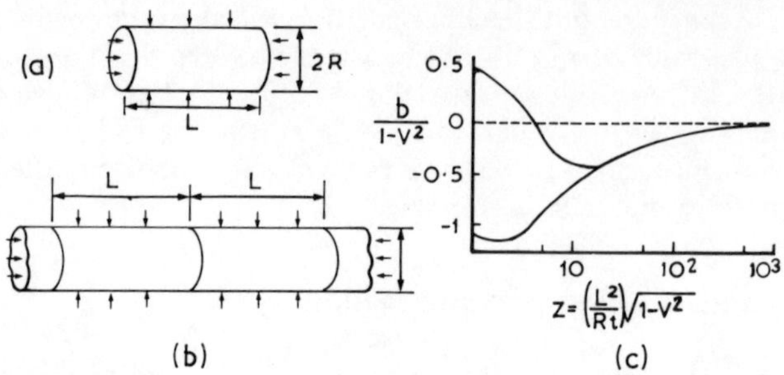

Fig. 6

and is associated with multiple, highly unstable post-buckling modes. The coupling between these modes together with the random nature of the distribution of initial imperfections in a practical shell leads to observed buckling pressures well below that given by equation (3). The imperfection-sensitivity of the pressure loaded spherical shell would appear to be as severe as that for the axially loaded cylinder [8]. In fact the variation of P_s/P_c versus $\bar{\delta}/t$ for the imperfect sphere is almost coincident with that obtained for the cylinder (Fig.3).

7 Clamped spherical cap under uniform pressure

This shell-loading combination is different from any of those considered so far in that the initial loading path is markedly *nonlinear* owing to the bending influence from the clamped edge. The geometry of the cap can be defined by two parameters λ and ϕ which, in the notation of Fig.7, are given by

$$\lambda = 2\,.\,\sqrt[4]{3(1-\nu^2)}\,.\,\sqrt{\frac{H}{t}} \quad \text{and} \quad \phi = \frac{H}{a}\,.$$

A description of the complex buckling phenomena associated with the spherical cap is best accomplished by defining specific ranges of these two parameters.

(a) $\phi \gtrless 0.125$, $\lambda \gtrless 3.5$. In this range the cap exhibits nonlinear stiffening behaviour (Fig.8(a)). The deformations are symmetrical thoughout.

(b) $\phi \gtrless 0.125$, $3.5 \gtrless \lambda \gtrless 5.5$. The shell deforms symmetrically and becomes unstable by reaching a limit point (Fig.8(b)). Under force loading the cap would then snap to a highly deformed symmetrical configuration.

(c) $\phi \gtrless 0.04$ (very shallow), $\lambda > 5.5$. After following a symmetrical initial loading path the perfect cap bifurcates into an asymmetrical mode which is of the form (Fig.7),

$$W_{\text{asymmetrical}} = W(r)\cos n\theta \qquad n = 2, 3, 4\,\ldots \tag{4}$$

The critical load at which bifurcation occurs is roughly constant at about 0.77 P_c where P_c is defined in equation (3). The stability of the asymmetrical modes has been determined by Fitch [9] who modified the initial post-buckling theory of Koiter to take into consideration the initial nonlinear

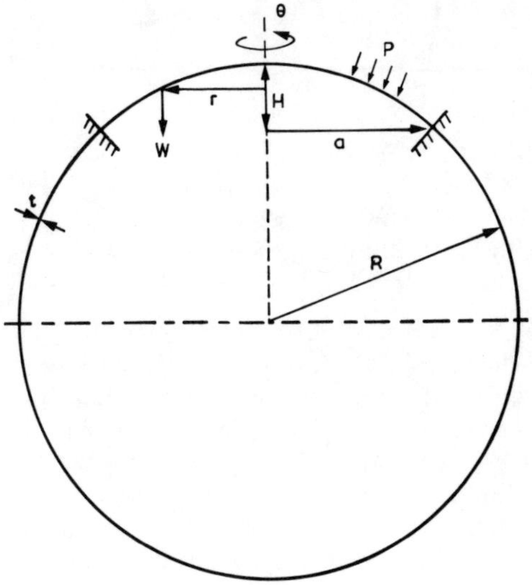

Fig. 7

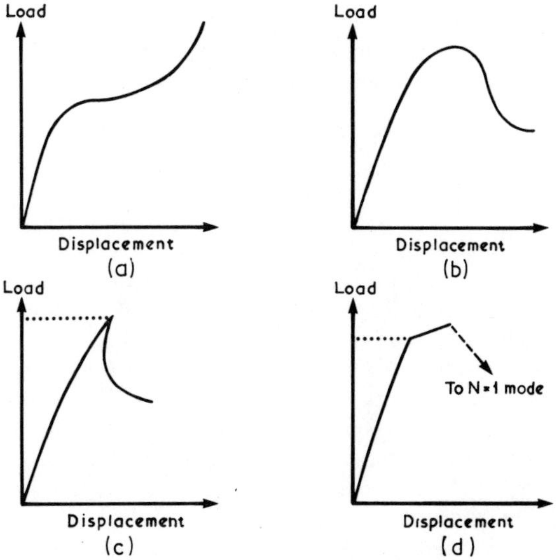

Fig. 8

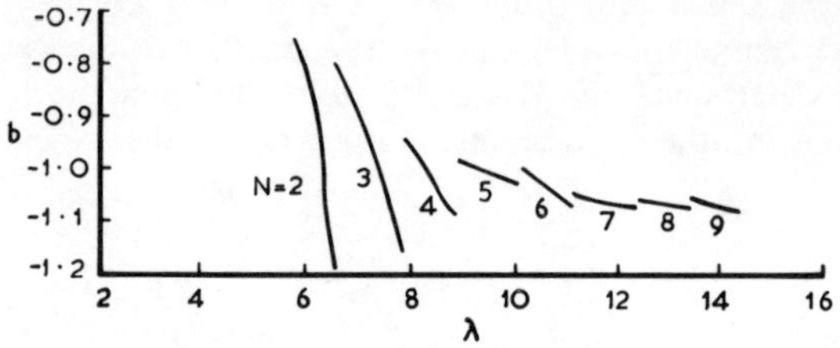

Fig. 9

loading path of this shell. His results for the coefficient b are shown in Fig.9 from which it is clear they are all unstable and the cap will consequently be highly imperfection-sensitive (Fig.8(c)).

(d) $0.04 \gtrsim \phi <?$, $\lambda \lesssim 5.5$. From recent experimental tests [10, 11] it has been determined that the stability of the asymmetrical modes described above shows a dependence on ϕ. If this parameter is greater than about 0.04 the asymmetrical modes having $n \geqslant 2$ will be initially stable. However the asymmetrical mode $n = 1$, consisting of a single edge dimple, is a possible but highly unstable configuration for the cap, although it is never an instigator of any instability (c.f. equation 4). The coupling between this mode and those having $n \geqslant 2$, together with the random imperfections in practical shells (which probably precipitate the coupling) nullifies the seemingly great practical potential of the shells in this range (Fig.8(d)).

(e) $\phi \to 1$ (Hemisphere). As the cap tends to a hemispherical form the influence of the edge bending disturbance becomes confined to a narrow edge layer and most of the shell tends to remain in a membrane state of stress. The post-buckling characteristics consequently tend to those displayed by the complete sphere.

8 Discussion

The behaviour of the various shell-loading combinations discussed illustrate many of the buckling phenomena associated with thin shells. In nearly all of the cases considered the initial post-buckling behaviour of the perfect shell was unstable and, in fact, this is a characteristic of most of the shell-loading combinations of practical interest. A knowledge of the imperfection-sensitivity of a shell is thus of paramount importance for design. The Koiter initial post-buckling theory has enabled some qualitative assessment of the degree of imperfection-sensitivity of some simple shell types to be made. It cannot be used to predict the buckling load of a practical shell however since, in general, the imperfection distribution will be of a random nature. Correlating the buckling strength of imperfect shells with a statistical description of the initial imperfections would appear to be the best approach to this problem. This however remains to be done. With the lack of theoretical guidance most thin shell design is carried out at the present time by using empirical relationships derived from a statistical analysis of a large quantity of experimental data (a collection of these relationships has recently been published [12]). This approach would appear to be the only realistic one for many of the complex thin shell structures encountered in modern practice and is likely to remain so for some time yet.

References

1 Koiter, W. T. 'Elastic stability and post-buckling behaviour', in Langer, R. E. (Ed.) *Nonlinear Problems*, University of Wisconsin Press (1963).

2 Budiansky, B. 'Post-buckling behaviour of cylinders in torsion', *Proceedings of the second IUTAM symposium on the theory of thin shells,* Copenhagen (Sep. 1967).

3 Koiter, W. T. 'The effect of axisymmetric imperfections on the buckling of cylindrical shells under axial compression', *Proceedings of the Koninklijke Nederlandse Akademie van Wetenschappen,* Amsterdam, Series B (1963).

4 Hutchinson, J. W. 'Buckling and initial post-buckling behaviour of oval cylindrical shells under axial compression', *Journal of Applied Mechanics* (Mar. 1968), pp.66–72.

5 Kempner, J. and Chen, Y. N. 'Large deflections of an axially compressed oval cylindrical shell', *Proceedings of the 11th International Congress of Applied Mechanics,* Berlin (1964).

6 Batdorf, S. B. 'A simplified method of elastic stability analysis for thin cylindrical shells', NACA Report 874 (1947).

7 Budiansky, B. and Amazigo, J. C. 'Initial post-buckling behaviour of cylindrical shells under external pressure', *Journal of Mathematics and Physics,* vol.47 (Sep. 1968), no.3.

8 Hutchinson, J. W. 'Imperfection sensitivity of externally pressurised spherical shells', *Journal of Applied Mechanics* (Mar. 1967), pp.49–55.

9 Fitch, J. R. 'The buckling and post-buckling behaviour of spherical caps under axisymmetric load', Report SM29, Division of Engineering and Applied Physics, Harvard University (Dec. 1968).

10 Tillman, S. C. 'On the buckling behaviour of shallow spherical caps under a uniform pressure load', *International Journal of Solids and Structures,* vol.6 (1970).

11 Tillman, S. C. 'An experimental investigation of the buckling behaviour of shallow spherical shells', *RILEM International Symposium Proceedings,* Buenos Aires, Argentina (Sep. 1971).

12 Baker, E. H., Kovolesky, L. and Rish, F. L. *Structural Analysis of Shells,* McGraw-Hill (1972).

7 Geometric nonlinearity and stability

J. F. Dickie*

1 Introduction

Structural nonlinearity has been widely studied recently; Oden [6] surveys 180 selected references concerning both geometric and material nonlinearity. Stricklin *et al.* [8] update this review in a recent very relevant paper.

In a number of structural forms under applied load, the change of geometry in the structure becomes significant and the structure whilst remaining elastic exhibits a nonlinear load-deflection relationship. Where this results in a loss of stiffness the possibility of total collapse, attributable to a local or general loss of stiffness must be investigated.

The toggle bar and arch shown in Fig.1 both illustrate this decreasing stiffness and as one would expect behave rather similarly. Reference [9] concerns the toggle bar and is of particular significance. A typical load-deflection curve is given in Fig.2 and concerns a symmetrical deflected configuration, although the possibility of a bifurcation occurring into a transitional antisymmetrical mode is possible, the associated onset of buckling being rather more sudden than that associated with the gradual loss of stiffness leading to P_{cr}. Nevertheless as the majority of problems confronting the engineer are unsymmetrical, observation of this decay in stiffness enables the critical load to be rapidly ascertained.

The shallow arch is discussed at length to demonstrate stability criteria and also as an example of the relative complexity of an apparently simple problem. A finite element solution is outlined. Exact solutions are compared with approximate series solutions and also discrete element solutions. A number of problems are briefly discussed to demonstrate the application of this discrete element approach. They are confined to considerations prior to buckling.

2 The shallow arch

The behaviour of shallow arches has been the subject of many investigations. Nonlinear closed form solutions and latterly discrete solutions have been widely pursued. In the first instance the arch attracted attention as one of the few practical nonlinear problems readily amenable to solution; subsequently the available closed

* Simon Engineering Laboratories, University of Manchester.

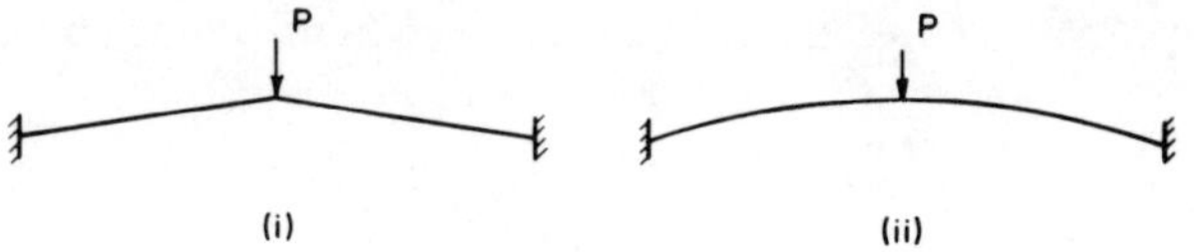

Fig. 1

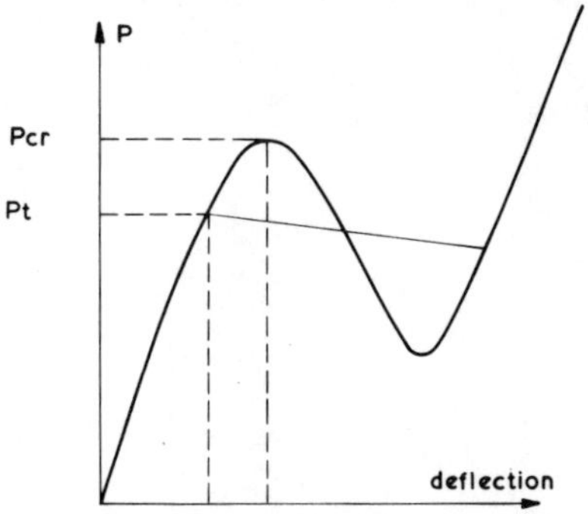

Fig. 2

form solutions have been employed to verify more general numerical approaches. Additionally the arch portrays many of the stability phenomena associated with more complex structural forms.

Consider the shallow arch shown in Fig.3 together with the coordinate system. The axial strain due to deformation is given by

$$\epsilon = \{\partial u/\partial a - \omega\}/R^2 + \{\partial \omega/\partial a\}^2 2R^2 \qquad (1)$$

where ϵ is the axial strain in the arch and u and ω the tangential and radial displacements respectively. The change in curvature is

$$\kappa = \{\partial^2 \omega/\partial a^2\}/R^2. \qquad (2)$$

The total energy of the system in non-dimensional form is

$$H = \tfrac{1}{2} \int_0^{\beta} \epsilon^2 \mathrm{d}a + (t^2/24R^4) \int_0^{\beta} \{\partial^2\omega/\partial a^2\}^2 \mathrm{d}a - \phi(W, \omega)/EtfR \qquad (3)$$

with $\phi(W, \omega)$ dependent upon the manner of loading considered.

For tangential equilibrium, the variation of H with respect to u must be zero; thus

$$\partial\epsilon/\partial a = 0 \quad \text{and} \quad \epsilon = C \qquad (4)$$

with C being proportional to the compressive membrane force. Equation (4) may be written as

$$\epsilon = -t^2\rho^2/12R^2. \tag{5}$$

For radial equilibrium, the variation of H with respect to ω must be zero. Using (1) and (5) and an applicable loading term in (3) leads to a linear differential equation for the particular loading case being considered.

3 Uniform radial loading

$$\phi(W, \omega) = \int_0^\beta (W\omega/\beta)\mathrm{d}a$$

and the differential equation is

$$\omega'''' + \omega'' = q/\mu^2 \tag{6}$$

in which the prime denotes differentiation with respect to η, $\eta = \rho a$, $\mu = \rho\beta$ and

$$q = 12R^3\beta^3W/Et^3f\mu^2 - R\beta^2.$$

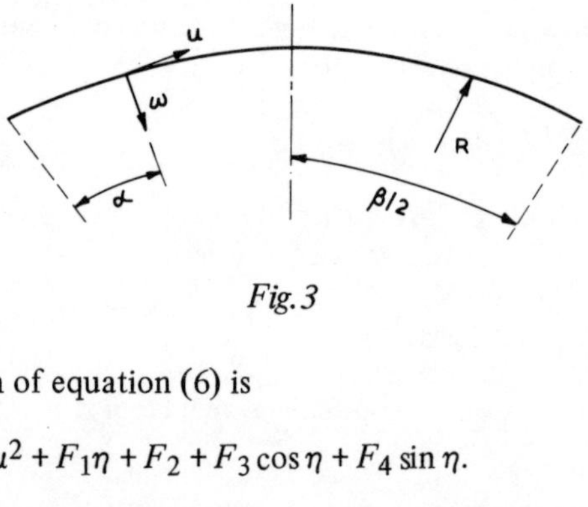

Fig. 3

The general solution of equation (6) is

$$\omega = q(\eta^2 - 2)/2\mu^2 + F_1\eta + F_2 + F_3\cos\eta + F_4\sin\eta. \tag{7}$$

Substitution of the particular geometrical and equilibrium boundary conditions gives a set of four simultaneous equations in terms of F_1, F_2, F_3, F_4 and q in terms of μ. An additional equation may be obtained from equations (1) and (6):

$$(1/\mu)\int_0^\mu \epsilon d\eta = -t^2\rho^2/12R^2. \tag{8}$$

Thus

$$-\int_0^\beta \omega d\eta + (\mu^2/2R\beta^2) \int_0^\beta (\omega')^2 d\eta + t^2\mu^3/12R\beta^2 = 0.$$

This equation is common to all the types of loading considered.

4 Centre point vertical loading

$$\phi(W, \omega) = W\omega\{\delta(a - \beta/2)\}$$

in which $\delta(a - \beta/2)$ = the Dirac function. The differential equation is

$$\omega'''' + \omega'' = \{12WR^3\beta^3/Et^3f\mu \,.\, \delta(\eta - \mu/2) - R\beta^2\}/\mu^2. \tag{9}$$

The general solution to equation (9) may be written as

$$\omega = \{F_1 + F_2\eta + F_3(1 - \cos\eta) + F_4(\eta - \cos\eta)$$
$$+ (12WR^3\beta^3/Et^2f\mu)(\eta - \mu/2 - \sin(\eta - \mu/2))\, U(\eta - \mu/2)$$
$$- R\beta^2(\eta^2/2 - 1 + \cos\eta)\}, \tag{9a}$$

in which

$$U(\eta - \mu/2) = 0; \quad \eta < \mu/2$$
$$= 1; \quad \eta > \mu/2.$$

Using the applicable boundary conditions and the relevant general solution in conjunction with equation (8) the solution to a particular problem may be obtained.

The load deflection curve is defined in both pre- and post-buckle regions as illustrated in Fig.2.

For an unsymmetrical problem the solution yields the only possible equilibrium path. For a symmetrical problem the possibility of an antisymmetrical mode occurring must be considered. Comparison of deflections determines whether this transitional mode occurs in the stable region of the load deflection curve and hence governs the stability criteria. If the transitional mode occurs in the unstable region the symmetrical buckling mode is obviously the applicable criteria, although the arch passes into a transitional equilibrium path after it has buckled along a symmetrical equilibrium path.

Fig.4 illustrates a typical curve for buckling load factors and concerns a pinned arch of rectangular section thickness t, breadth f, subjected to centre point loading. For $4.5 < \lambda < 12$ symmetrical buckling occurs, for $9.2 < \lambda < 12$ an antisymmetrical

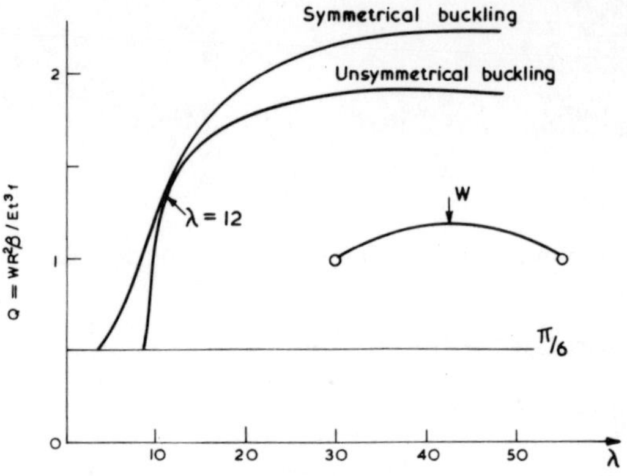

Fig. 4

buckling path exists but in the unstable region and for $\lambda > 12$ antisymmetrical buckling occurs.

5 Approximate solution

Solutions may be obtained using a series solution. Assuming the radial deflection at any point on the arch to be

$$\omega = \sum^{\infty} A_n \omega_n(\zeta),$$

equation (3) becomes

$$H\lambda/\beta^5 = \lambda/2 \left\{ \int_0^1 \left(\sum^{\infty} a_n \omega_n(\zeta) \right) d\zeta - \tfrac{1}{2} \int_0^1 \left(\sum^{\infty} a_n \,.\, \partial\omega_n(\zeta)/\partial\zeta \right)^2 d\zeta \right\}^2$$

$$+ (1/24\lambda) \int_0^1 \left(\sum^{\infty} a_n \,.\, \partial^2\omega_n(\zeta)/\partial\zeta^2 \right)^2 d\zeta - \phi(W^*, \omega) \tag{10}$$

in which $\zeta = a/\beta$; $\lambda = \beta^2 R/t$, a geometrical arch parameter; $a_n = A_n/R\beta^2$; and $W^* = WR/Et^2 f\beta$.

The series chosen to represent the deflected shape depends upon the end conditions of the arch and preferably the series should satisfy all geometric and equilibrium boundary terms as a smaller number of terms then suffice. Suitable series are as follows. For both ends pinned:

$$\omega = \sum^{\infty} A_n \sin n\pi\zeta \tag{11}$$

and for one pinned end and one clamped end:

$$\omega = \sum^{\infty} A_n \sin n\pi\zeta \cos(\pi\zeta/2). \tag{12}$$

For both ends clamped

$$\omega = \sum^{\infty} A_n \sin n\pi\zeta \sin \pi\zeta. \tag{13a}$$

The expansion

$$\omega = \tfrac{1}{2} \sum^{\infty} A_n(\cos(n-1)\pi\zeta - \cos(n+1)\pi\zeta) \tag{13b}$$

proves convenient in this case.

For any position of equilibrium,

$$\partial H^*/\partial a_n = 0. \tag{14}$$

Using equation (10) and equation (14) yields a set of n simultaneous nonlinear algebraic equations in $a_1 \ldots a_n$ and W. Repeated solution, using the Newton-Raphson method, of these equations in terms of any particular unknown gives the load-deflection curve. In the case of the pinned arch satisfactory agreement is obtained between the series solution and the exact solution by taking a four-term series ($n = 1, 2, 3, 4$).

For arches possessing one pinned end and one clamped end a three-term series suffices. For the clamped arch a six-term series is normally necessary. A number of results are compared in Table 1. The analysis may be extended to incorporate imperfection considerations as in reference [7].

Although the solutions presented have been obtained relatively straightforwardly a number of limitations are apparent. The assumptions require radial loading, resulting calculations are lengthy and the structural form is specifically defined. It is however a 'line' structure and may be considered as a two-point boundary value problem, a category that also includes columns and towers. Reference [3] employs a simple approach to examine a number of post-buckling problems concerning unguyed towers and stacks of various cross-section.

The approach has been adopted by the author to examine the tower problem schematically illustrated in Fig.5(a). Each element is assumed to remain straight.

The slope changes computed from the curvature of each element are assumed to take place at prescribed nodes. Assuming values of the indeterminate base moment and base shear nodal positions may be determined using elementary theory and successively moving up the column.

Table 1. Comparative buckling loads

Boundary condition		Closed form solution Symmet-rical	Closed form solution Transi-tional	Series solution Symmet-rical	Series solution Transi-tional	Matrix method Symmet-rical	Matrix method Transi-tional
Pinned	73	9.4	8.2	9.4	8.2	9.6	8.0
Pinned	55	22.1	19.2	22.1	19.2	22.5	18.0
Clamped	73	10.2	9.3	10.4	9.3	10.3	9.0
Clamped	55	23.9	20.7	24.1	20.7	24.1	20.4

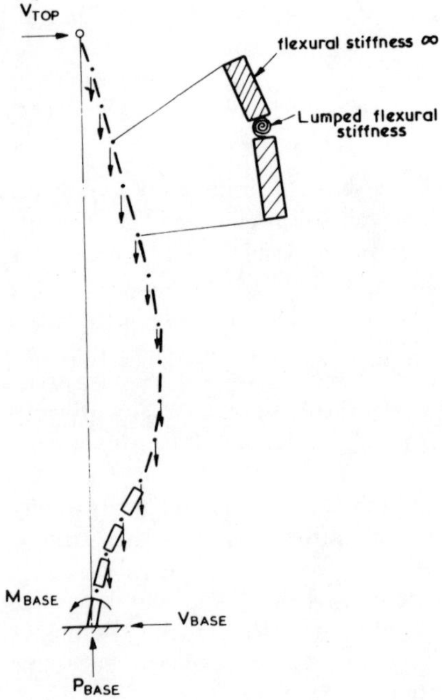

Fig.5(a) Deflected configuration of tower.

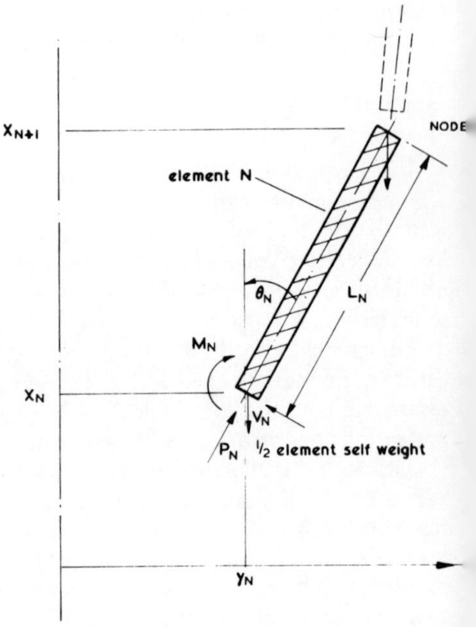

Fig.5(b) Nodal loading.

Consider the element shown in Fig.5(b). The slope change along element N, a_N, is given by

$$a_N = V_N L_N^2 / 2EI_N + M_N L_N / EI_N$$

$$L_N = L(1 - P_N / EA_N) \tag{15}$$

(self weight is incorporated in nodal loading)
whence

$$x_{N+1} = x_N + L_N \cos\theta_N; \quad \theta_{N+1} = \theta_N - a_N$$

$$y_{N+1} = y_N + L_N \sin\theta_N. \tag{16}$$

If the base moment and base shear have been assigned their correct values, successively applying equations (15) and (16) would lead to zero moment and zero displacement at the top of the tower. Normally an iterative procedure is necessary updating M_{base} and V_{base} using Newton iteration. Although useful in examining line structures the limitations are clear and to examine complex structures it is necessary to proceed along more general lines as subsequently described. As previously demonstrated the instability of slender elastic arches is amenable to closed form solution and a number of approximate methods; the effect of pre-stressing may also be included. Problems of interest to the practising engineer as well as the theoretical researcher are still to be investigated. Previous solutions tend to be limited in application as they are based on particular geometric configurations as well as loading conditions.

In order to develop a general method of arch analysis a numerical discretization technique must be used. Although the discrete method already described is useful for structural forms related to two-point boundary value problems a completely generalized approach is applicable to a wide range of problems. The method outlined here is fully described in reference [2].

Consider a line member with reference axes P, Q and R; P is the chord connecting member ends, Q and R are coincident with principal axes of the member section. It is required to relate joint forces $\mathbf{F}$ and joint displacements Δ as defined in Fig.6; it is convenient to consider force and displacement vectors $\mathbf{R}$, $\mathbf{U}$, $\mathbf{P}$ and $\mathbf{E}$ as defined in Figs 7 and 8.

Relating these vectors

$$\mathbf{U} = \mathbf{T} \times \Delta \tag{17}$$

$$\mathbf{F} = \mathbf{T}' \times \mathbf{R}. \tag{18}$$

The transformation $\mathbf{T}$ derives from direction cosines of member axes and both $\mathbf{T}$ and its transpose $\mathbf{T}'$ are valid for cases of gross deformation as $\mathbf{U}$ and Δ refer to original coordinate positions.

$$F = \{F_{AX}, \cdots\cdots M_{BZ}\}$$
$$\Delta = \{X_A \cdots\cdots \theta_{BZ}\}$$

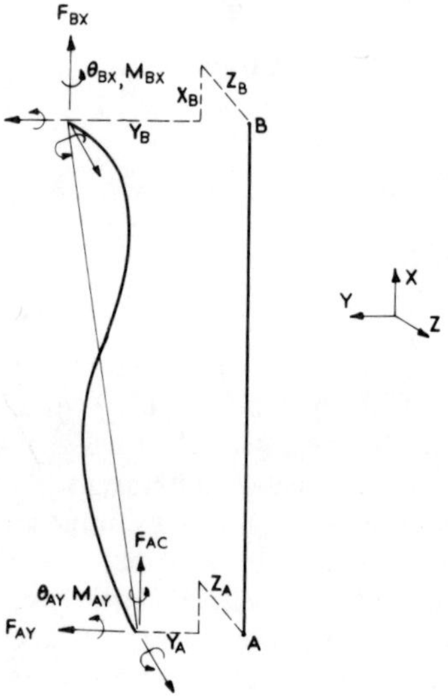

Fig. 6 Joint forces and displacements.

The relationship between the displacements **U** and **E** is somewhat more complex as **E** refers to the deformed member coordinate positions. Taking

$$\mathbf{E} = \mathbf{A} \times \mathbf{U} \tag{19}$$

to determine **A** it is necessary to consider

(i) the direction cosines $l_p, \ldots n_r$, of the deformed member axes relative to the undeformed member axes;

(ii) the angles of sway, defined as β_q the angle between the P' axis and the projection of the P axis onto the Q' plane, and β_r the angle between the P' axis and the projection of the P axis onto the R' plane.

Equation (19) takes the form

$$\epsilon = ((L + u)^2 + v^2 + w^2)^{1/2} - L$$

$$\phi_{ABr} = l_{r'}\theta_{ABp} + m_{r'}\theta_{ABq} + n_{r'}\theta_{ABr} - \beta_r$$

etc. (20)

$$R = \{R, S_q, M_{ABr}, M_{BAr}, M_{ABp}, M_{BAp}, S_r, M_{ABq}, M_{BAq}\}$$

$$U = \{U, v, \theta_{ABr}, \theta_{BAr}, \theta_{ABp}, \theta_{BAp}, W, \theta_{ABq}, \theta_{BAq}\}$$

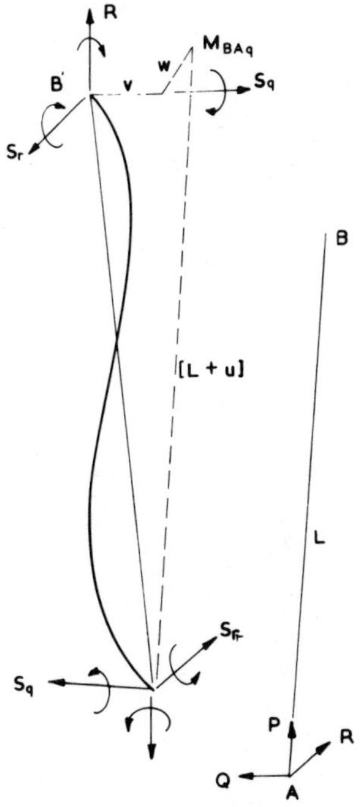

Fig. 7

Defining incremental displacements

$$\mathbf{x} = \{\delta x_A, \ldots \delta_{BZ}\}$$

$$\mathbf{u} = \{\delta u, \ldots \delta_{BAq}\}$$

$$\mathbf{e} = \{\delta \epsilon, \delta \phi_{ABr}, \ldots \delta \phi_{BAq}\} \tag{21}$$

and rewriting (17) and (18) for incremental displacements gives

$$\mathbf{u} = \mathbf{T} \times \mathbf{x} \tag{22}$$

$$\mathbf{e} = \mathbf{a} \times \mathbf{u}. \tag{23}$$

The transformation **a** is obtained by partially differentiating (20) with respect to each of the intermediate displacement variables in turn. Consideration of the virtual

$$P = \{f, M_{ABr}, M_{BAr}, T, M_{ABq}, M_{BAq}\}$$

$$E = \{\epsilon, \phi_{ABr}, \phi_{BAr}, \alpha, \phi_{ABq}, \phi_{BAq}\}$$

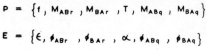

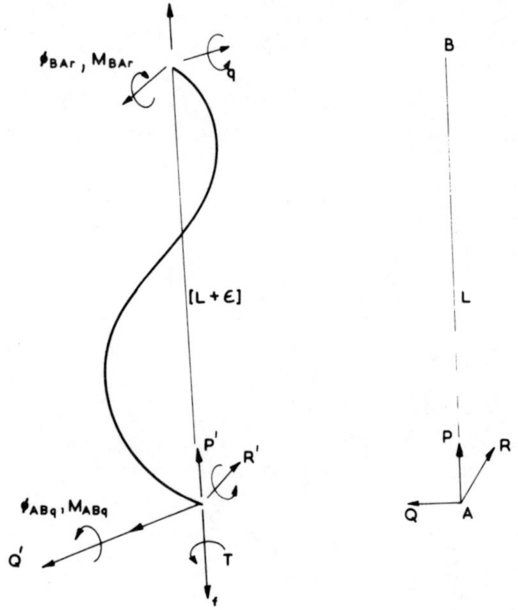

Fig. 8

work equation gives

$$\mathbf{R}' \times \mathbf{u} = \mathbf{P}' \times \mathbf{e}$$

hence

$$\mathbf{R} = \mathbf{a}' \times \mathbf{P} \tag{24}$$

similarly defining a set of incremental force vectors **f**, **r** and **p** with

$$\mathbf{f} = \mathbf{T}' \times \mathbf{r}. \tag{25}$$

The transformation from incremental intermediate forces to incremental basic forces is obtained by partially differentiating (8) with respect to each of the intermediate displacement variables and each basic force variable in turn;

$$\mathbf{r} = \mathbf{a}' \times \mathbf{p} + \mathbf{d} \times \mathbf{u}. \tag{26}$$

In order to obtain **K** as defined

$$\mathbf{P} = \mathbf{K} \times \mathbf{E} \tag{27}$$

it is necessary to assume displacement functions for the member; cubic polynomials suffice. Standard energy considerations, including axial shortening due to bowing and twisting, enable the total energy of the system to be obtained. It may be assumed that member twist varies linearly along the length of the member. Partial differentiation of the total energy gives the member stiffness characteristics in the form

$$f = EA((\epsilon/L) + (2\phi_{ABr}^2 - \phi_{ABr} \cdot \phi_{BAr} + 2\phi_{BAr}^2 + 2\phi_{ABq}^2$$

$$- \phi_{ABq} \cdot \phi_{BAq} + 2\phi_{BAq}^2)/30 + (I_q + I_r)a^2/2AL^2)$$

$$M_{ABr} = \phi_{ABr}((4EI_r/L) + (4fL/30)) + \phi_{BAr}((2EI_r/L)$$

$$- (fL/30)) + EI_{qr}(4\phi_{ABq} + 2\phi_{BAq})/L$$

etc. (28)

Partial differentiation of (28) with respect to each of the member basic displacement variables in turn enables the member incremental basic stiffness matrix **k** to be obtained with

$$\mathbf{p} = \mathbf{k} \times \mathbf{e}. \tag{29}$$

Combining (22), (23), (25), (26) and (27) gives incremental forces in terms of incremental displacements.

$$\mathbf{f} = \mathbf{T}'(\mathbf{a}' \times \mathbf{k} \times \mathbf{a} + \mathbf{d})\mathbf{T} \times \mathbf{x} \tag{30}$$

or

$$\mathbf{f} = \mathbf{K}_2 \times \mathbf{x}. \tag{31}$$

Internal joint forces are obtained from (18), (25) and (27)

$$\mathbf{F} = \mathbf{T}' \times \mathbf{a}' \times \mathbf{K} \times \mathbf{E} \tag{32}$$

or

$$\mathbf{F} = \mathbf{K}_1 \times \Delta. \tag{33}$$

Equation (33) contains the nonlinear stiffness equations of a rigid jointed member in space; equation (30) defines the tangent stiffness equations. The non-

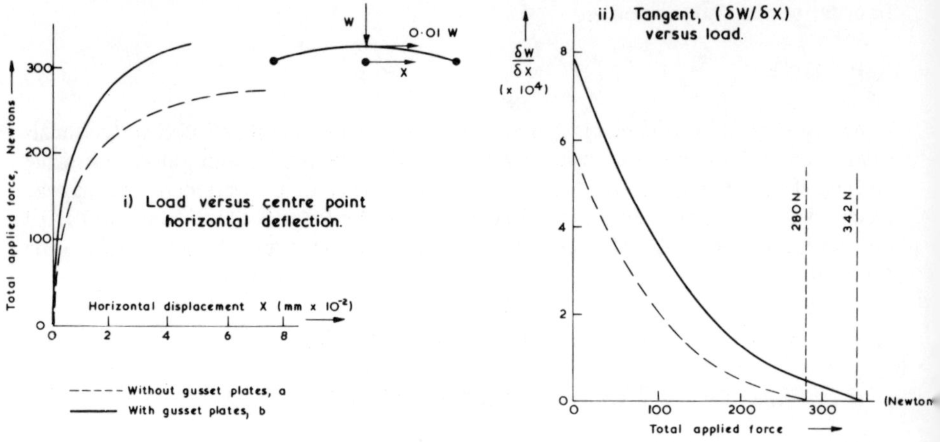

Fig. 9

linear stiffness equations are assembled in an identical manner to that employed in linear analysis by summing component contributions. The resulting equations are solved in a piecewise linear manner using the Newton-Raphson method. The application of these equations to three-dimensional cases is described in reference [2] together with three-dimensional pin-jointed structures of which cable structures are a particular case.

Alternative matrix formulations abound in the literature. The described analysis was primarily developed for analysing structures comprising slender prismatic members. Analytical results have been extensively compared with experimental results obtained from both basic and complex structural forms. Agreement has always been satisfactory for considerable nonlinearities (>100%). Predicted buckling loads have also been accurate.

Other authors' solutions have been compared wherever possible and other than with reference [9] no serious discrepancy has occurred.

As a comparison between alternative arch analyses Table 1 shows comparative buckling loads, relating to centrally loaded arches, obtained using: (1) closed form solutions; (2) approximate series solution; and (3) the matrix approach. For the purposes of the discrete element analysis each arch was subdivided into eight equal segments. The buckling load was ascertained by observing load stiffness relationships at particularly relevant degrees of freedom. The order of 10 not necessarily equal load increments suffices.

Transitional buckling loads in symmetrical cases may be determined by providing a small antisymmetrical disturbance as illustrated in Fig.9. Whilst the critical load may be determined from the load deflection curve the author has found curves plotted as in Fig.9 (ii) allow a more rapid determination of buckling loads. It is convenient to observe the tangent to the appropriate load displacement characteristic as this vanishes at the point of instability. These curves are computed as opposed to a graphical interpretation of the load-displacement characteristic. The curves presented in Fig.9 concern experiment in addition to theory. The gusset plates referred to in this figure were attached for jointing purposes and although relatively small had a significant effect on the buckling load.

The portal frame shown in Fig.10 was analysed in reference [4] as an example illustrating two-dimensional frame analysis including change of geometry. For the symmetrical loading shown the true buckling load is $W = 950$ lb. This compares to that predicted by an elastic eigenvalue buckling load of $W = 2550$ lb corresponding to the second mode.

The three dimensional structure shown in Fig.11 illustrates an additional complication. Although in nonlinear cases superposition is not permissible and the quadrant symmetry is not of practical significance quite frequently the design loading criteria require only one half of the structure to be analysed. This is achieved by applying restraints at nodes, for example if the loading were symmetrical about the x-axis at nodes, 14, 8, 2, 1, 5, 11, 17 then

$$\delta_y, \theta_x, \theta_z = 0.$$

In this particular case lateral buckling of the rib members occurs and this would be precluded by symmetrical restraints.

The case illustrated in Fig.12(a) derives from a dome structure of span 100 m, and is indicative of the size and cost of the procedures previously described. Assuming symmetry about a diameter could be employed the solution would require approximately 200 K locations in direct storage. As the arch ribs are the main structural members some lumping of the ring members is permissible radically reducing the problem size. The reduced problem shown in Fig.12(b) costs approximately £30/iterative cycle and up to 6 cycles/load may be required.

In conclusion although the numerical techniques described are now widely used they are dependent on what is essentially a modelling procedure and it is emphasized that mathematical models are just as susceptible to error as experimental ones.

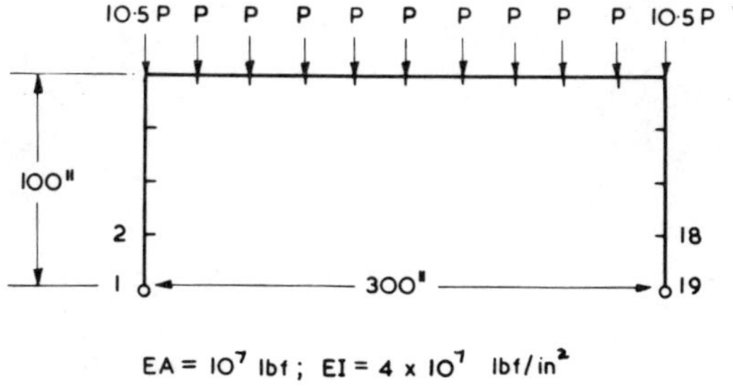

Fig.10

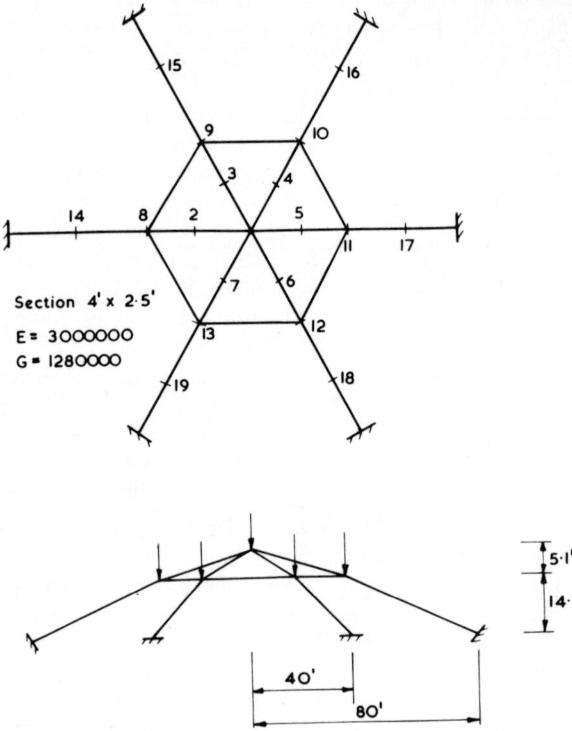
15
16
9
10
3
4
14
8
2
5
11
17
7
6
Section 4' x 2·5'
13
12
E= 3000000
G= 1280000
19
18
5·1'
14·9'
40'
80'

Fig. 11

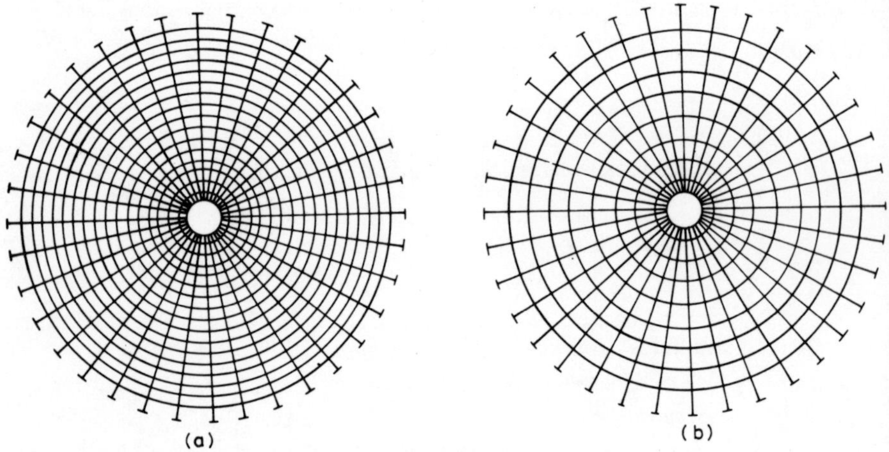
(a)
(b)

Fig. 12

References

1 Dickie, J. F. and Broughton, P. 'Stability criteria for shallow arches', *Journal of the Engineering Mechanics Division, ASCE,* vol.97 (1971), no.EM3, pp.951–965.

2 Dickie, J. F. and Broughton, P. 'Geometric nonlinearity', *J. Struct. Mech.,* vol.1 (1972), no.2, pp.249.

3 Harrison, H. B. 'Nonlinear elastic analysis of unguyed towers and stacks', to be published in *Journal of the Engineering Mechanics Division, ASCE.*

4 Jennings, A. 'Frame analysis including change of geometry', *Journal of the Structural Division, ASCE,* vol.94 (1968), no.ST3.

5 Livesley, R. K. *Matrix Methods in Structural Analysis.* Pergamon Press, New York (1964).

6 Oden, J. T. 'Finite element applications in nonlinear structural analysis'. *Proc. Symp. on application of finite element methods in Civil Engineering,* ASCE, Vanderbilt Univ., 1969.

7 Schreyer, H. L. 'The effect of initial imperfections on the buckling load of shallow circular arches', *Journal of Applied Mechanics Trans. ASME.*

8 Stricklin et al. 'Solution procedures for structural analysis', *AIAA* (Mar. 1973).

9 Tezcan, S. S. and Mahapatra, B. C. 'Tangent stiffness matrix for space frame members', *Journal of the Structural Division, ASCE,* vol.95 (1969), no.ST6.

10 Williams, F. W. 'An approach to the non-linear behaviour of the members of a rigid jointed plane framework with finite deflections', *Quarterly Journal of Mechanics and Applied Mathematics,* vol.17 (1964), pp.451–469.

Appendix: general references

1 Ashwell, D. G., Sabir, A. B. and Roberts, T. M. 'Further studies in the application of curved finite elements to circular arches', *International Journal of Mechanical Sciences,* vol.13 (1971), pp.133–139.

2 Augilar, R. J. 'Snap-through buckling of framed triangular domes', *Journal of the Structural Division, ASCE,* vol.93 (1967), no.ST2, Proc. Paper 5202, pp.301–317.

3 Connor, J. J., Logcher, R. D. and Chan, S. C. 'Nonlinear analysis of elastic framed structures', *Journal of the Structural Division, ASCE,* vol.94 (1968), no.ST6, Proc. Paper 6011, pp.1525–1547.

4 Conway, H. D. and Lo, C. F. 'Further studies on the elastic stability of curved beams', *International Journal of Mechanical Science,* vol.9 (1967), no.10, pp.707–718.

5 Dickie, J. F. and Broughton, P. 'Stability considerations in shallow vaults and shallow domes'. *Pacific symposium on tension and space frames.* International Association of Shell Structures, Tokyo 1971.

6 Dickie, J. F. and Broughton, P. 'Shallow circular vaults', *Journal of the Engineering Mechanics Division, ASCE,* vol.99 (1973), no.EM1.

7 Ginnini, M. and Miles, G. A. 'A curved element approximation in the analysis of axi-symmetric thin shells', *International Journal for Numerical Methods in Engineering,* vol.2 (1970), no.3, pp.459–476.

8 Gjelsvik, A. and Bodner, S. R. 'Energy criterion and snap buckling of arches', *Journal of the Engineering Mechanics Division, ASCE,* vol.88 (1962), no.EM5, Proc. Paper 3304.

9 Haisler, W. E., Stricklin, J. A. and Stebbins, F. J. 'Development of evaluation of solution procedures for geometrically nonlinear structural analysis', *AIAA Journal,* vol.10 (1972), no.3, pp.264–272.

10 Lind, N. C. 'Numerical buckling analysis of arches and rings', *Structural Engineer,* vol.44 (1966), no.7, pp.245–248.

11 Livesley, R. K. *'Matrix methods of structural analysis'.* Pergamon, Oxford (1964).

12 Martin, H. C. 'Finite element formulation of geometrically nonlinear problems', Paper US 2-2, *Proceedings of Japan-US Seminar on Matrix in Structural Analysis and Design*, Tokyo, 1969, pp.1–53.

13 Oran, C. 'Complementary energy method for buckling of arches', *Journal of the Engineering Mechanics Division, ASCE,* vol.94 (1968), no.EM2, Proc. Paper 5116, pp.639–651.

14 Oran, C. and Reagan, R. S. 'Buckling of uniformly compressed circular arches', *Journal of the Engineering Mechanics Division, ASCE,* vol.95 (1969), no.EM4, Proc. Paper 6732, pp.879–895.

15 Richard, R. M. and Blacklock, J. R. 'Finite element analysis of inelastic-structures', *AIAA Journal,* vol.7 (1969), no.3.

16 Stricklin, J. A., Haisler, W. E. and Von Riesemann, W. A. 'Geometrically nonlinear structural analysis by the direct stiffness method', *Transactions of the ASCE, Journal of the Structural Division,* vol.97 (1971), no.ST9, pp.2299–2314.

17 Walker, A. C. 'A nonlinear finite element analysis of shallow circular arches', *International Journal of Solids and Structures,* vol.5 (1969), pp.97–107.

18 Wempner, G. A. and Ewbank, T. 'Buckling of circular arches under non-uniform pressure', *Journal of the Engineering Mechanics Division, ASCE,* vol.89 (1963), no.EM4, Proc. Paper 3597, pp.17–20.

19 Wright, D. T. 'Membrane forces and buckling in reticulated shells', *J. Struct. Div. Am. Soc. Civ. Engrs,* vol.91 (1965), no.ST1, pp.173–201.

8 Numerical post-buckling analysis

J. W. Butterworth*

1 Introduction

The complexity of elastic post-buckling behaviour of many practical structures frequently results in the associated numerical analysis assuming formidable proportions and accounts for many past attempts to simplify the problem. Fortunately the expanding body of knowledge on the general theory, by indicating possible behavioural patterns, provides the analyst with invaluable prior knowledge of the many pitfalls that await him.

A common technique used to construct equilibrium paths for discrete elastic systems under single-parameter loading is to evaluate equilibrium configurations corresponding to varying levels of applied loading thereby establishing a sequence of points in load-configuration space. When a sufficient number of such points have been established in a region of interest their locus forms the required equilibrium path. We describe here an approach to the problem of generating such a sequence of equilibrium configurations.

2 Analysis under prescribed loading

We consider first the case of analysis in which the load magnitude Λ is prescribed and the configuration which will equilibrate this load is sought. It is assumed that the system possesses a total potential energy function $V(Q_i, \Lambda)$ where the Q_i are a set of n generalized coordinates and Λ is a load parameter. V is also assumed to be linear in Λ such that

$$V(Q_i, \Lambda) = U(Q_i) - \Lambda Q_i w_i$$

where $U(Q_i)$ denotes the strain energy of the system, $\Lambda Q_i w_i$ defines the potential energy of the loading and w_i is a vector of (constant) external loads (summation over repeated subscripts is assumed).

Suppose that Q_i defines a configuration which does not satisfy equilibrium and which is to be modified in order to approach more closely the true equilibrium configuration. Let δQ_i denote an increment vector such that $(Q_i + \delta Q_i)$ is a better

* Department of Civil Engineering, University of Surrey.

approximation to the desired equilibrium configuration. The necessary condition for equilibrium is

$$V_i = 0, \quad i = 1, 2, 3 \ldots n,$$

where the subscript on V denotes differentiation with respect to the corresponding generalized coordinate.

Expanding V_i in a Taylor series about the Q_i configuration

$$V_i(Q_i + \delta Q_i) = V_i(Q_i) + V_{ij}(Q_i)\delta Q_j + 1/2 V_{ijk}(Q_i)\delta Q_j \delta Q_k + \ldots$$

If $(Q_i + \delta Q_i)$ is to be an equilibrium state it is necessary that

$$V_i(Q_i + \delta Q_i) = 0.$$

Dropping the higher order terms in the series expansion and accepting that $(Q_i + \delta Q_i)$ will now only be an improved rather than an exact equilibrium state we obtain

$$0 = V_i + V_{ij}\delta Q_j + \ldots \text{ higher order terms dropped,}$$

where all derivatives are evaluated at the point Q_i.

Since V is linear in Λ it follows that

$$V_i = U_i - \Lambda w_i$$

and

$$V_{ij} = U_{ij}.$$

Substituting into the truncated series expansion yields

$$U_{ij}\delta Q_j = \Lambda w_i - U_i. \tag{1}$$

Since U_i (by Castigliano's theorem) defines the force exerted by the structure in the Q_i direction it follows that $(\Lambda w_i - U_i)$ is a vector representing the extent to which equilibrium is satisfied in each of the Q_i directions. U_{ij} is the well-known tangent stiffness matrix which defines the relationship between incremental loads and the corresponding incremental displacements.

The set of linear simultaneous equations may be solved for the δQ and an improved approximation $(Q + \delta Q)$ obtained. Taking $(Q + \delta Q)$ as a new starting point the process may be repeated and the solution further improved. Fig.1 illustrates schematically the solution procedure for a two-degree-of-freedom system with a prescribed load of Λ_1. (0,0) is taken as the initial approximation and successively improved configurations are obtained as the points a, b, c . . . etc.

This (Newton-Raphson) procedure, provided that the first approximation is sufficiently close to the solution and that the equilibrium path is well behaved (rising monotonically with load), will usually converge to the required solution, but requires considerable numerical computation in the formulation and reduction of U_{ij} at each step. It has been shown [1] that greater numerical efficiency can often be obtained by carrying out a less sophisticated correction to the configuration and simply repeating the correction cycle a greater number of times. To obtain further points on the equilibrium path the load is increased and the latest configuration further modified to achieve equilibrium under the new prescribed load.

3 Stability of equilibrium

Having located an equilibrium configuration by means of the analysis just described or some other means (see Section 4) it is desirable, and in general necessary, to establish whether or not the equilibrium is stable. For example, if a fundamental equilibrium path loses its stability at a point of bifurcation it is obviously necessary to detect this and not continue to follow a new unstable path. The properties of the tangent stiffness matrix U_{ij}, evaluated at the equilibrium state in question, provide a convenient means for checking stability. When $V(Q_i, \Lambda)$ is linear in Λ it has been shown that

$$U_{ij} = V_{ij}.$$

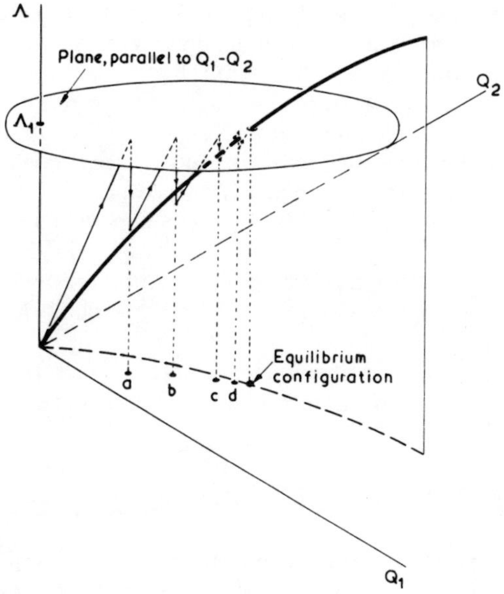

Fig. 1

V_{ij} is of course the matrix of the quadratic form defining the second variation

$$\delta^2 V = V_{ij}\delta Q_i \delta Q_j \quad \text{(see Chapter 2)},$$

and must have the same rank, index and signature as the A_{ij} and C_i matrices (see Chapter 2) as these quantities are all invariant under non-singular linear transformations.

It has been shown how the C_i coefficients determine stability and in view of the above relationship between C_i and U_{ij} all the necessary information on stability can be deduced from the U_{ij} matrix which is usually the most convenient to work with numerically. The number of positive, negative and zero C_i will have a one-to-one correspondence with the number of positive, negative and zero pivots encountered in a routine elimination or decomposition process on the U_{ij} matrix, so that if U_{ij} is formed at the equilibrium state of interest a simple examination of its properties will reveal the state of all the corresponding stability coefficients.

4 Analysis under prescribed displacement

Maintaining a fixed load value and seeking the corresponding equilibrium configuration as described in Section 2 will break down in certain circumstances and prove inconvenient in others. For example, if the prescribed load is very near or above a local snap-buckling load or if there are multivalued equilibrium states in the same vicinity at the given load level. In such cases a different approach is indicated and one method that has been used is to prescribe a displacement rather than a load, thereby defining a hyper plane in load configuration space, and then to seek the intersection of the equilibrium path with this hyper plane by varying the remaining $(n - 1)$ generalized coordinates and the load Λ until equilibrium is achieved. The existence of an intersection point will naturally depend upon which generalized coordinate is prescribed and the value imposed – the criteria governing this choice are discussed later. An iterative scheme is derived in a similar manner to the controlled load case, with slight changes to accommodate the now variable load parameter Λ.

Let Q_r be the generalized coordinate that is to be assigned a fixed value of $\overline{Q}_r$, and let (Q_i, Λ) be an approximation to the required equilibrium state (Q_i, $i \neq r$, all variable; $Q_r = \overline{Q}_r$).

Let $(\delta Q_i, \delta\Lambda)$ denote a set of increments $(\delta Q_r = 0)$ such that $(Q_i + \delta Q_i, \Lambda + \delta\Lambda)$ is a better approximation to the required equilibrium state. Expanding $V_i(Q_i, \Lambda)$ about the point (Q_i, Λ) in a Taylor series gives

$$V_i(Q_i + \delta Q_i, \Lambda + \delta\Lambda) = V_i(Q_i, \Lambda) + V_{ij}(Q_i, \Lambda)\delta Q_j + V_i'(Q_i, \Lambda)\delta\Lambda + \dots,$$

where primes denote differentiation with respect to Λ. Setting $V_i(Q_i + \delta Q_i, \Lambda + \delta\Lambda) = 0$ and dropping the higher order terms in order to take a linear step towards the desired equilibrium state yields

$$V_i(Q_i, \Lambda) + V_{ij}(Q_i, \Lambda)\delta Q_j + V_i'(Q_i, \Lambda)\delta\Lambda = 0.$$

With the form of V defined earlier it follows that

$$V' = U' - Q_i w_i,$$

$$V_i' = -w_i,$$

$$V_i = U_i - \Lambda w_i,$$

$$V_{ij} = U_{ij}$$

and after substituting,

$$U_{ij}\delta Q_j = \Lambda w_i - U_i + \delta\Lambda w_i \qquad (2)$$

where all evaluations are assumed to be at the point (Q_i, Λ).

This system of simultaneous equations (2) involves the $n + 1$ unknowns δQ_i, $i = 1, 2 \ldots n$, and $\delta\Lambda$, and may be solved with the additional equation

$$\delta Q_r = 0.$$

After solving for $(\delta Q_i, \delta\Lambda)$, the improved approximation $(Q_i + \delta Q_i, \Lambda + \delta\Lambda)$ may be used as the starting point for a further improvement cycle, and so on until equilibrium is satisfied (as evidenced by $\Lambda w_i - U_i$ tending to zero).

The process of solution described above is illustrated schematically in Fig.2. The course of the solution cannot be represented precisely by points in the $\Lambda - Q_i$ space because the intermediate loadings are not definable by the single load parameter Λ.

5 Choice of prescribed load or displacement

The preceding Sections 2 and 4 describe how equilibrium states may be found by a simple iterative process during which either the load or one displacement component is temporarily held constant. Which of these two methods to employ depends on the nature of the equilibrium path in the region of the analysis. Generally a controlled load level is employed when tracing the initial part of a fundamental equilibrium path emerging from the unloaded state, since this region of the path is likely to be well-behaved and progress along it well defined by Λ. Also, the analyst is more likely to have knowledge of the order of load appropriate to a structure rather than the order and direction of displacements.

Fig.3 shows four equilibrium states established at loads of Λ_1, Λ_2, Λ_3 and Λ_4 on the equilibrium path of a system which reaches a limit point at a load Λ_{cr}. An attempt to obtain a fifth point at a load of Λ_5 would result in the iterative solution procedure failing to converge. Fig.4 shows three (stable) equilibrium states estab-

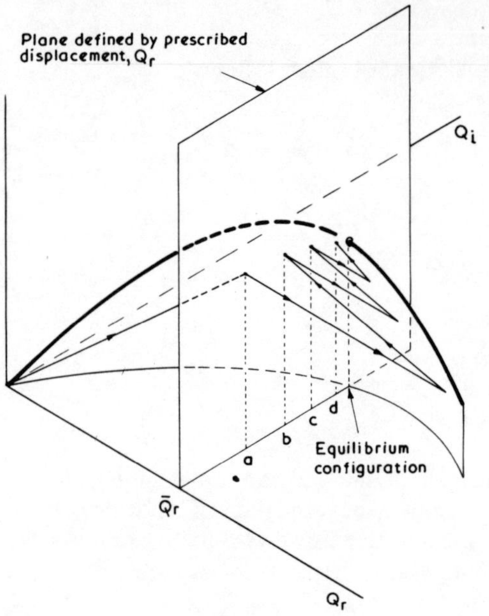
Plane defined by prescribed displacement, Q_r
Λ
Q_i
a
b
c
d
Equilibrium configuration
$\bar{Q}_r$
Q_r

Fig. 2

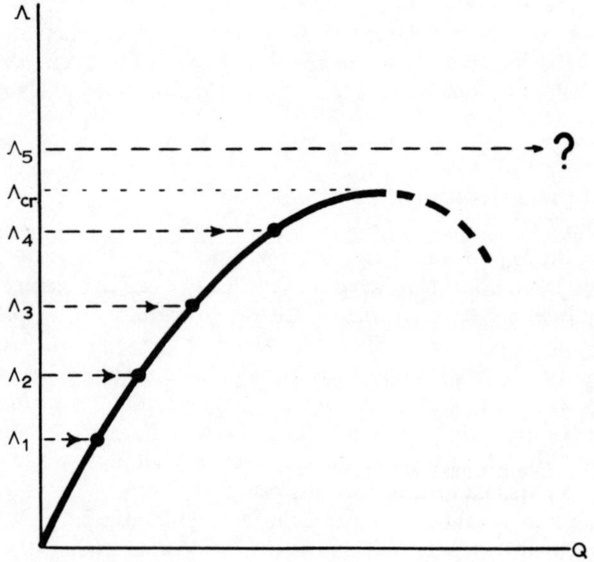
Λ
Λ_5
?
Λ_{cr}
Λ_4
Λ_3
Λ_2
Λ_1
Q

Fig. 3

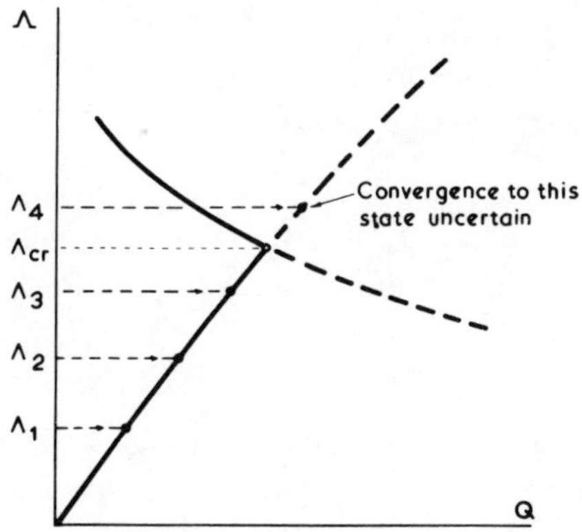

Fig.4

lished at loads of Λ_1, Λ_2 and Λ_3 on the fundamental path of a bifurcating system. At a load level of Λ_4 an (unstable) equilibrium state has been found on the unstable portion of the fundamental path above the critical point. In both these cases analysis under controlled loading is satisfactory only for loads less than Λ_{cr} and difficulty is likely to be experienced as the load approaches Λ_{cr}.

To study post-buckling behaviour it is generally necessary to abandon the simple load control method and adopt the more flexible displacement control technique. Two questions arise at this point – when should displacement control be adopted, and which displacement component should be controlled? If the analyst has prior knowledge of the expected behaviour of a structure he may opt for displacement control throughout the analysis – for example, in the case of a shallow arch or dome in which it is known that snap buckling in a particular mode is likely to occur it could be advantageous to control the displacement component that suffers the greatest change during buckling. Fig.5 illustrates a very simple case of this kind in which vertical displacements of Q_1, Q_2, Q_3 . . . are imposed on the two-member arch and the corresponding points on the equilibrium path obtained.

In general it must be assumed that no detailed prior knowledge of the shape of the equilibrium path is available in which case displacement control will be required when one of two possibilities occurs: (a) a limit point is reached; (b) a point of bifurcation is reached.

The limit point is the more general of the two possibilities – either as a 'natural' limit point or as a case of an imperfect bifurcating system – and is considered first. As the critical load at the limit point is approached the convergence rate of the iterative process will deteriorate and above the critical load the process will diverge. At the same time any displacement components which have a significant projection in the direction of buckling will begin to acquire increasingly large values. Displacement control should be implemented at this point and obviously the best component to control is the one which has the greatest projection on the tangent to the equilibrium path at the critical point since this will best define

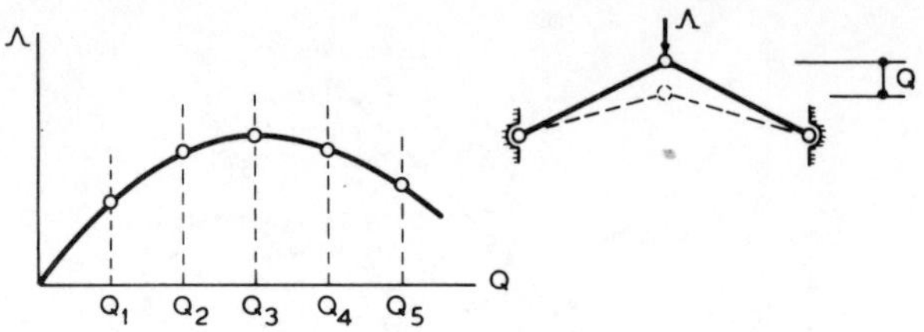

Fig.5

progress along the path. The tangent to the equilibrium path at the critical point is given by x_j, the eigenvector corresponding to the eigenvalue which vanishes at the critical point [2]. x_j will thus be given by the solution to

$$U_{ij}(Q^{\mathrm{cr}}, \Lambda^{\mathrm{cr}})x_j = 0,$$

where $(Q^{\mathrm{cr}}, \Lambda^{\mathrm{cr}})$ denotes the critical point.

The critical point is not usually precisely known numerically and so either an approximate eigenvector must be calculated or alternatively the direction of the equilibrium path can be estimated from the last two or three points established on it. If the Q_i coordinates are in dimensionless form (numerically this is desirable), then the largest element of the eigenvector indicates the appropriate displacement to control. The magnitude and sense of the displacement that is then prescribed may be taken as a suitable proportion of the current value of the selected displacement. Fig.6 illustrates a snap-buckling equilibrium path on which four points have been established under load control at loads of Λ_1, Λ_2, Λ_3 and Λ_4. Control has then been transferred to the displacement mode with Q_r selected as the direction having the greatest projection on the known direction of the equilibrium path at Λ_4. Four further points are then established by prescribing the sequence of displacements Q_r^1, Q_r^2, Q_r^3 and Q_r^4. The critical point $(Q_r^{\mathrm{cr}}, \Lambda^{\mathrm{cr}})$ is passed without difficulty.

In the case of highly contorted paths it may prove necessary to change the mode of control more than once; for example in the case of a path which loops back upon itself [5] (see Chapter 3, Fig.1.).

The second possibility requiring displacement control is the occurrence of a bifurcation point. The discussion is limited for the moment to the case of a distinct bifurcation in which only a single stability coefficient changes sign and only one post-buckling path exists. The direction of the post-buckling path is again determined by x_j, the eigenvector corresponding to the eigenvalue vanishing at the critical point, and the appropriate displacement to control is the one which has the greatest projection on x_j. However, it is no longer sufficient to simply prescribe the selected displacement Q_r and assume that an equilibrium state on the post-buckling path will automatically be found. Fig.7 shows four equilibrium states on the prebuckling path of a bifurcating system obtained under load control. If the displacement $\overline{Q}_r$ was then prescribed it can be seen that there are two possible equilibrium

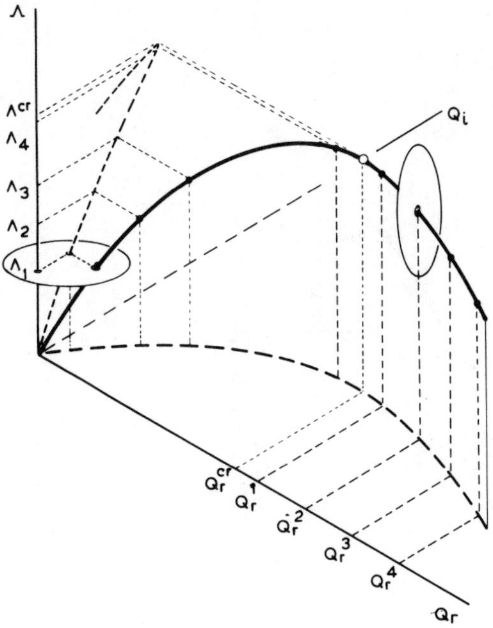

Fig. 6

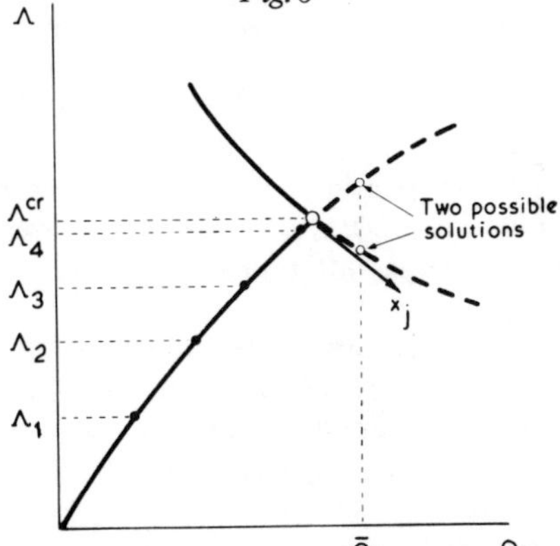

Fig. 7

states, one on the post-buckling path and one on the unstable portion of the fundamental path, and there is no guarantee as to which of these two would be converged upon by the iterative analysis. To ensure convergence to the post-buckling path it is necessary to ensure that the *first approximation* used in the analysis is *sufficiently close* to the post-buckling path. This may be accomplished by changing all the Q_i in such a way that the configuration moves *along the eigenvector* to take up a starting position well clear of the fundamental path and close to the post-

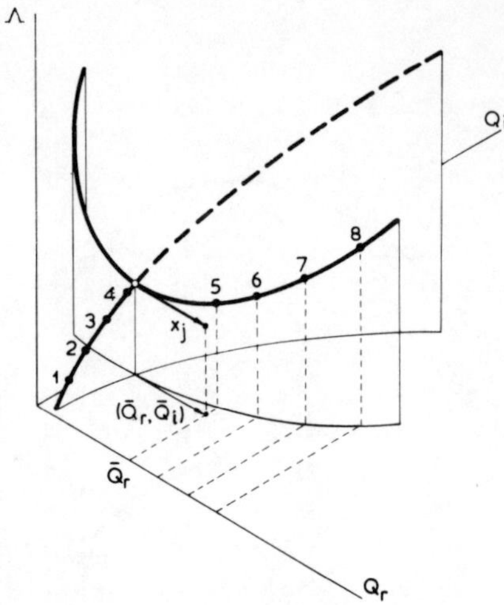

Fig. 8

buckling path. In other words a proportion a of the eigenvector is simply added to every displacement of the system as shown below

$$\overline{Q}_i = Q_i^{\mathrm{cr}} \pm a x_i,$$

where $\overline{Q}_i$ denotes the initial configuration from which the attempt to converge on to the post-buckling path is to be made. During the subsequent iterative analysis only Q_r will be held constant at a value Q_r, the remaining Q_i being allowed to vary along with the load parameter Λ.

Fig.8 illustrates the above procedure schematically. Points 1, 2, 3 and 4 have been found using controlled load levels, $(\overline{Q}_r, \overline{Q}_i)$ denotes the first approximation to a point on the post-buckling path from which the analysis will converge to the point 5. Points 6, 7 and 8 are then obtained by incrementing the controlled displacement Q_r.

Having explored one branch of the post-buckling path by using a starting point of

$$\overline{Q}_i = Q_i^{\mathrm{cr}} + a x_i,$$

it is an easy matter to return to the critical point and follow the other branch by using a starting point of

$$\overline{Q}_i = Q_i^{\mathrm{cr}} - a x_i,$$

which results in taking a first step in the opposite sense along the same eigenvector.

The case of compound branching points [3], in which two or more stability coefficients vanish simultaneously leading to multiple post-buckling path generation, cannot be treated as simply as the distinct critical points. The simultaneous vanishing of several eigenvalues means that there is no longer a unique direction defined by an eigenvector at the critical point – there will exist instead an r-dimensional eigen-space (in the case of r simultaneously vanishing eigenvalues). A systematic search of this eigen-space would be necessary in order to locate all the possible post-buckling paths, of which there may be very many [3]. Of course, in practice it is quite likely that imperfections, both physical and numerical, will cause the analysis to follow an imperfect path, missing completely the 'perfect' compound branching point. The problem is studied in detail by Thompson and Hunt [4].

6 Concluding remarks

Two simple nonlinear analysis schemes have been described based on the well known Newton-Raphson procedure and utilizing either prescribed load or displacement. Analysis under load control generally proves satisfactory for pre-buckling behaviour provided the nonlinearity is not too great. Displacement control provides a means for carrying out analysis of the post-buckling behaviour. Stability of an equilibrium state is easily established by an inspection of the properties of the tangent stiffness matrix U_{ij} used in the nonlinear analysis.

References

1 Supple, W. J. and Vine, G. B. Research Report No.102, Space Structures Research Centre, University of Surrey (1973).

2 Thompson, J. M. T. 'A general theory for the equilibrium and stability of discrete conservative systems', *Z. angew. Math. Phys.*, vol.20 (1969), pp.797–846.

3 Johns, K. C. and Chilver, A. H. 'Multiple path generation at coincident branching points', *Int. J. mech. Sci.*, vol.13 (1971), pp.899–910.

4 Thompson, J. M. T. and Hunt, G. W. 'A theory for the numerical analysis of compound branching', *Z. angew. Math. Phys.*, vol.22 (1971), pp.1001–1015.

5 Chilver, A. H. 'Coupled modes of elastic buckling', *J. Mech. Phys. Solids*, vol.15 (1967), pp.15–28.

9 Concluding remarks

The contents of the previous chapters have attempted to introduce the reader to the fundamentals of post-buckling behaviour of elastic structures and to venture into an examination of these properties in some detail. At first sight the material may appear somewhat formidable in that the equations may seem very complicated and the figures intricate. However, once the mathematical symbolism used is understood and the pictorial representation of results is carefully studied much of the apparent complication should disappear. The results of the general branching theory will be seen to be elegant and all-embracing.

The concept of a *bifurcation point* with reference to equilibrium paths is of crucial importance to the study of structural instability even though such a point is somewhat of an abstraction in practical structural terms. By this we mean that if we determine an experimental load-deflection curve for a structure then we shall invariably witness a limit point type of behaviour owing to the inherent structural imperfections. However, the work of Roorda described in Chapter 4 demonstrates how by artificially varying the initial imperfections we may study the change in the limit point behaviour in the vicinity of a point of bifurcation. The so-called perfect system when viewed as one of a family of imperfect systems is seen to represent a *singular* situation and the bifurcation point is a *singular point.* The generation of bifurcation points in the load-deformational response of discrete structural systems (i.e. systems whose deformation is represented for analysis purposes by a set of discrete components) is demonstrated in Chapters 2 and 3 in general terms. The theory presented is valid for the *initial* post-buckling regime but the range of validity may be extended formally by merely retaining and evaluating higher order energy coefficients in the analysis. However, experimental evidence for frame buckling and analytical results for plate and shell buckling described in Chapters 4, 5 and 6 respectively indicate that initial post-buckling theory is adequate for relatively large buckling displacements and explains all the observed phenomena.

There are many instances where elastic instability may be encountered in structural design and analysis and it has not been the intention here to present and itemize all such occurrences, but instead to present the underlying fundamentals which govern all elastic buckling situations. In any particular analysis of elastic instability we wish to determine the critical load of a structure, that is, the maximum load the structure may carry without buckling. We have seen from the results presented here that in order to obtain the correct value of this load for a practical structure we must guarantee that all of the appropriate information is fed into the analysis. For instance, we must not fail to include the effects of inevitable imperfections in geometry or loading arrangement. Further, all the probable buckling modes must be examined particularly with regard to any deleterious coupling which may occur. We may mention, for instance, the possible interaction between a local and global mode of instability. Again, if a numerical procedure is to be used it must be capable of dealing with multiple equilibrium paths if these are present in the problem and also must not fail to detect bifurcation points on fundamental equilibrium paths. The method should also be able to cope with locally high nonlinearity of equilibrium paths.